PHYSICS of the ENVIRONMENT

by

Herbert Inhaber
Carleton University
Ottawa, Canada

ANN ARBOR SCIENCE
PUBLISHERS INC
P.O. BOX 1425 • ANN ARBOR, MICH. 48106

> UNIVERSITY OF
> STRATHCLYDE
> DAMAGED
> 181289

Copyright © 1978 by Ann Arbor Science Publishers, Inc.
230 Collingwood, P.O. Box 1425, Ann Arbor, Michigan 48106

Library of Congress Catalog Card No. 77-76911
ISBN 0-250-40187-8

Manufactured in the United States of America
All Rights Reserved

D
628.5015'3
INH

PREFACE

Is physics useful? This may seem a strange question with which to begin a physics book, but it does reflect some people's attitude. Not every viewpoint can be changed by one book—but I will try to influence one group—college students who are not physics majors.

Interest in physics can be generated not by rehashing the principles that launched a thousand books, but by presenting them in a new light. To many students, physics is connected with the disturbing and disoriented society they see around them—the pollution in the air and water, the bombs and missiles, the billions spent on grandiose engineering projects while social needs are left unmet. To these students, physics is associated with the problems of, not the solutions for, society.

If we are ever to control any of the menaces facing us, we must learn more about them. Prime among these menaces is environmental degradation. The application of physics principles has caused a fair proportion of the decreased environmental quality. For example, if we knew nothing about heat or thermodynamics, we certainly could not have built an internal combustion engine, and air pollution would be definitely lessened. If hunters did not have an intuitive feel for the dynamics of a bullet, there would be fewer endangered species. But the genie is out of the bottle; we do know these principles and there is no forgetting them. We can use these principles to our advantage, however, to measure the degree of pollution—for measurement is crucial to understanding—and to control it. Though almost no one sits down to solve environmental problems by starting from the basic principles of physics, many do use these principles, consciously or unconsciously, when searching for solutions.

We would like eventually to control the environment by eliminating pollution. Before we can do this, we have to measure its state; that is, find out exactly how much pollution of different types exists. Though there has been much more research done on measurement than on control, we tend to be more interested in the latter. Since measurement generally costs much less than control, it is emphasized by environmental agencies. This book will reflect the preponderance of measurement research, though control of environmental problems will also be given much consideration.

We want to know not only the thoughts of physics but how physicists think. For example, consider the subject of approximations. Most people regard physics as an exact science, and it is. To get this degree of exactness, however, approximations must be made in almost every experiment and calculation. Some of the approximations are involved, but many are the "back of the envelope" variety. If a physicist wants to know whether a small error by a measuring apparatus could affect the result of an experiment, the so-called order of magnitude of the results should be calculable within a short time. Since approximation cannot be taught, we will try to illustrate the concept whenever possible throughout the book.

One use of this book will be in the undergraduate courses which are often entitled "environmental studies." In some colleges and universities, these courses have reached the status of a full-fledged department. In most others, the courses are a melange gathered from a number of departments, including chemistry, biology and engineering. Physics has usually been conspicuous by its absence. But if we are going to make a concerted attack on the environmental problems which confront us, we must utilize all our scientific and technical knowledge. Physics can make significant contributions.

In this book, calculus is not used, algebra is kept to a simple level and physics principles are carefully explained. With these points in mind, this book would be suitable for the "liberal arts" physics course given in many colleges and universities. These courses are designed to give nonscience majors some idea of what physics is about, without going into the detail of more advanced courses. The aim of this book, then, is not to teach physics from the ground up. The object here is to illustrate some of its uses.

Physics can have unpleasant side-effects on our environment; the evidence is all too painful. But physics can also work to improve our environment, as it helps to measure its state and to control it. This book has been written with the idea of putting physics to work not for death, but for life.

For my parents

A scientific advisor for the Canadian Atomic Energy Control Board and Sessional Lecturer for Carleton University in the Department of Physics, Herbert Inhaber received his BS from McGill University, MS from the University of Illinois, did postgraduate work at the University of Rochester, and received his PhD in low-temperature physics from the University of Oklahoma. His present work involves analysis of the risk produced from many types of energy systems and teaching a self-designed course, "Physics of the Environment." His previous work at the Canadian Department of the Environment comprised formulation of environmental indices and other quality measures.

Dr. Inhaber has worked with U.S. Steel Applied Research Laboratories, Xerox Research Laboratories and the Science Council of Canada, and also has been a Visiting Lecturer at Yale University where he was jointly on the faculty of the Department of History of Science and Medicine and School of Forestry and Environmental Studies.

The author of a book, *Environmental Indices*, and more than 40 papers, Dr. Inhaber is a member of the Canadian Association of Physicists, Sigma Xi, Sigma Pi Sigma and is listed in *Who's Who in the East*.

CONTENTS

1. **WHAT IS PHYSICS OF THE ENVIRONMENT?** 1
   - Approach to Environmental Problems 1
     - Marine Pollution 1
     - Oil Spills on Water 3
   - Instrumentation 3
     - Luminescence and Absorption 3
     - The Ideal Instrument 6
     - Varying Sensitivities 7
     - The Cost of Detection 7
   - The Numbers of Environmental Physics 10
     - Carbon Dioxide and the Biosphere 10
     - How Accurate is the Measurement? 13
   - Summary 16
   - Problems 17
   - References 18

2. **CONSERVATION OF ENERGY, FORCE AND THE ENVIRONMENT** 21
   - Introduction 21
     - The Flywheel 22
     - Oil and Mechanics 25
   - Measurement 31
     - Damping Vibrations 31
   - Control 34
     - Sludge and Centripetal Force 34
     - Cool and Clear Water 36
     - Problems 39
     - References 40

3.	SOUND IN THE ENVIRONMENT	41
	Introduction	41
	The Decibel	41
	Limitations of the Decibel	43
	Adding Intensity Levels	46
	Highways and Noise	46
	Measurement	48
	Sound Level Meters	48
	River Velocity	49
	Dust in the Air	53
	Bacteria in Sludge	54
	Sonic Booms	56
	Control	56
	Walls as Insulators	56
	Openings in Insulators	59
	Summary	61
	Problems	62
	References	64
4	HEAT AND THERMODYNAMICS	67
	Introduction	67
	Smoke Density	68
	Thermal Mapping	70
	Measurement	73
	Thermal Accounting	73
	Power Plants and Weather	80
	Summary	82
	Problems	83
	References	85
5.	ELECTRECOLOGY: ELECTRICITY AND MAGNETISM	87
	Introduction	87
	Fish and Electricity	87
	Water Conductivity and Pollution	88
	Conductivity and Air Pollution	91
	Measurement	92
	Piezoelectricity and Air Pollution	94
	Acid Rain	98
	The Size of Air Pollutants	100
	Oil Slick Thickness	102
	Mining and Acid Wastes	105
	Microwaves on the Ocean	107

	Ocean Currents	109
	Polarography	111
	Control	113
	Filtering Water	113
	Electrostatic Filters	115
	Summary	117
	Problems	118
	References	122
6.	LIGHT AND OPTICS	125
	Introduction	125
	Light and Water	125
	Asbestos Detection	127
	Underwater Light Absorption	129
	Measurement	131
	Chlorophyll and Optics	131
	Oil and Light	136
	Oil Identification	137
	Quantifying Smell	140
	Remote Detection	144
	Sulfites and Sulfides	145
	Control	149
	Ultraviolet Light in Water	149
	Pesticides and Light	150
	Summary	150
	Problems	151
	References	156
7.	LASERS	157
	Introduction	157
	Laser Operation	157
	Measurement	161
	Particles in Dirty Water	161
	Oil and Lasers	163
	Air Pollutants	165
	Control	166
	Holograms of the Air	166
	Measuring Air Pollutants Directly	168
	Summary	170
	Problems	171
	References	172

8.	ONE ATOM, INDIVISIBLE: ATOMIC PHYSICS AND RADIOACTIVITY	173
	Introduction	173
	Radiation and Reactors	174
	Fallout, North and South	176
	Neutron Activation	178
	Water and Deuterium	181
	Measurement	184
	Half-Lives	184
	Sewage and Radioactivity	185
	Control	185
	Elimination of Bacteria	185
	Planning Sewage Systems	186
	Types of Oil	188
	Summary	190
	Problems	191
	References	193

ANSWERS TO SELECTED PROBLEMS 195

GLOSSARY 201

INDEX 221

CHAPTER 1

WHAT IS PHYSICS OF THE ENVIRONMENT?

This chapter is divided into three parts: how a physicist approaches an environmental problem, instrumentation and the use of numbers in the physics of the environment.

APPROACH TO ENVIRONMENTAL PROBLEMS

The approach to environmental and other problems differs from physicist to physicist. We will look first at the physical principles and quantities involved, then their measurement, and finally the results and conclusions.

Marine Pollution

The major causes of marine pollution have been categorized in Table 1.1. The left-hand row lists the sources of this pollution, and the middle vertical columns list its effects. By their nature, these two parts of the table have to be general in coverage.

The relative importance of each effect or source usually cannot be measured precisely. What *can* be measured is shown on the right-hand side of the table, where the physical quantities used to detect and control particular types of marine pollution are categorized. There are also some biological and chemical quantities which are used for the same purposes, but are not discussed here.

All of the quantities in the right-hand column are essentially physical measurements. Some, such as radioactivity and temperature measurement, fit easily into the established branches of physics; others, such as particle size and concentration of objects, can be found using more than one physics principle. Many of the physical measurements mentioned in the last column are discussed in later chapters.

Table 1.1 Major Categories of Water Pollution and Methods of Physical Measurement[1]

	Major Effects				Principal Sources					Physical Measurements
	Harm to Living Resources	Hazards to Human Health	Hindrance to Maritime Activities	Reduction of Amenities	Direct Outfalls and Rivers	Runoff from Land	Dumping from Ships	Seabed Mining Exploitation	Atmospheric Transfers	
Domestic Sewage	x	x		x	x	o	o			Temperature, color, turbidity
Inorganic Wastes	x	o			x		o		x	pH, specific conductance
Radioactive Materials		o			x		o		o	Radiation (alpha, beta and gamma counting)
Oil	o		o	x	x		o	o		Temperature, reflectivity
Petrochemicals and Organics	o	o			x		o			pH
Organic Wastes	x			o	x	o	o			pH
Heat	o				x					Temperature, reflectivity
Solid Objects	o		o	x	o		x			Particle size, concentration
Dredging Spoil, Inert Wastes	o		o	o	o		o	x		Temperature, color

Legend: x denotes important effect or source, o denotes significant effect.

Though quite general, Table 1.1 does show the wide variety of principles used to measure and control different types of pollution.

Oil Spills on Water

Perhaps Table 1.1 is oversimplified. An example of how a physicist would look more closely at a particular type of marine pollution—oil—is shown in Table 1.2. This table shows ways to detect crude oil on water after an oil spill using three branches of physics: mechanics, electromagnetism and optics. Some of the physical quantities have complicated units. These are not discussed in detail, since here the numbers alone are more important.

If the specific gravity, or density, were used to measure the presence of oil on seawater, a fairly precise instrument would be needed since the difference between the two substances is only about 15%. On the other hand, if the viscosity were employed, a greater contrast would exist. The viscosity of an average crude oil is greater than that of seawater by a factor of about 90 at 22°C and about 50 at 30°C.

A physicist interested in detecting oil spills might well construct a chart similar to Table 1.2. All other things being equal, the greatest interest would be in the physical properties which showed the largest contrast between oil and seawater.

In nature, all things are not always equal. When detecting oil spills there are uneven thicknesses of the oil layers, waves, oil-water froth stirred up by the waves and other conditions which make it difficult to measure the spills. Different types of environmental conditions demand different measurements.

INSTRUMENTATION

This leads to the problem of instrumentation. The many types of physics instruments preclude a complete discussion. However, a few aspects may be considered from the point of view of what instruments are suitable for which purposes, and why. The numbers contained in Table 1.2 said little about which properties of oil and seawater are detectable when, for example, there was a very small oil spill. In this case, the thinness of the oil might require different measurements from that of a large oil spill.

Luminescence and Absorption

Let us consider two approaches to instrumentation—the fluorescent and absorption methods. The two methods describe what happens when light

Table 1.2 Mechanical, Electrical and Optical Differences Between Oil and Seawater[2]
(at 25°C unless otherwise stated)

Physical Property	Specific Gravity at 22°C	Viscosity at 22°C (centistokes)	Viscosity at 30°C (centistokes)	Surface Tension at 22°C (dynes/cm)	Thermal Conductivity (cal/sec-cm-°C)	Relative Dielectric Constant (Ratio to Air)	Conductivity (millimhos/cm)	Index of Refraction (Visible Spectrum) (Ratio to Air)
Range for Oils	0.7-1.007	0.67-2970	0.59-562	20-37	$2.7\text{-}5.4 \times 10^{-4}$	1.85-2.94	Insulator	1.40-1.58
Range for Typical Crude Oils	0.79-0.93	4.08-255	2.76-113	24.4-31.8	$3.1\text{-}4.2 \times 10^{-4}$			
Average for Crude Oils	0.857	90	41	28	3.3×10^{-4}	2.0	Insulator	1.44
Seawaters	1.024	1.04	0.78	73.02	15×10^{-4}	7.8 at 1 MHz	16-60 (Conductor)	1.34

strikes an object. If the object is fluorescent, it gives off light, or radiation, after light hits it. The two types of light are usually not of the same frequency. The fluorescent lamp is the most common everyday example. Even in the most fluorescent substances, some light is absorbed and not radiated off. Although everything that fluoresces also absorbs to some extent, the two processes are different.

Figure 1.1 shows how they would work in measuring air pollutants. In the absorption method, shown in Figure 1.1a, air pollutant A is dissolved in water and chemical B is added to the solution. A reaction of the form A + B → C then occurs. The amount of chemical C present is then measured by passing light through the new solution, and noting how much light is absorbed using the meter M, which transmits a signal.

In Figure 1.1b, the fluorescent method of measuring A is shown. The pollutant A is mixed in the gaseous phase with gas B, and the reaction is

$$A + B \rightarrow C^*$$

where C* is a chemical which itself gives off light. To use a physical term, C* is excited. This means that some of the electrons in C* have more energy than usual. The type of light given off is measured by the meter M, and another signal is transmitted.

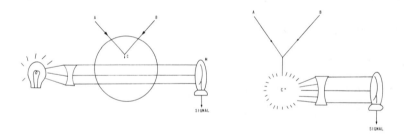

Figure 1.1 Measuring air pollutants by the (a) absorption and (b) luminescent methods.[3]

The sensitivities of the two processes will depend on the frequency that molecules of A collide with molecules of B. The molecules of A have to follow a circuitous route—gas to solution to combination—to be measured in the absorption process. As a result, the frequency of collisions is less than in the fluorescent process, in which A goes only from gas to combination with B. The fluorescent process is then inherently more sensitive. It may have other drawbacks—but one of its advantages is clear.

This may be put on a more quantitative basis by considering how the signal from meter M changes with the amount of pollutant A. In the absorption method, the intensity of light reaching the meter through the solution C generally decreases with increasing concentration of A. Figure 1.2a shows a typical nonlinear drop with concentration. In the fluorescent method, however, the change of intensity is directly proportional to the concentration, as shown in Figure 1.2b. This can be an advantage in interpreting results.

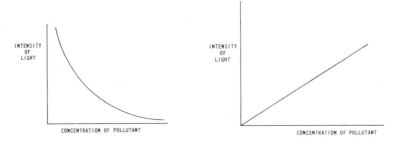

Figure 1.2 Response curves for the (a) absorption and (b) luminescent methods of measuring air pollutants.[3]

Depending on the range of concentration of pollutant, either of two competing methods of measurement may be better. To solve environmental problems, a combination of methods may be used to produce the most comprehensive answers.

The Ideal Instrument

The ideal instrument would be (1) highly specific, *i.e.,* measure only one pollutant at a time, (2) very sensitive, (3) able to give very reproducible results, (4) reasonably rapid and (5) preferably automated (to reduce cost and to simplify information processing). Each of these categories could be subdivided. For example, category 5 could include information on such points as the degree of automation and whether it includes sampling as well as a continuous output.

The perfect instrument, fulfilling these conditions, does not exist and probably never will. A picture of the state-of-the-art, as it pertains to some major water pollutants, is given in Table 1.3. In the first column of the table, "ions" refers to inorganic ions such as ammonium, sulfate and nitrate. "Other organics" generally refers to industrial wastes. The physics behind most of the methods of detection is explained in the Glossary. Adjectives like "good," "fair" and "poor" are to some extent subjective. However, it is useful to see the wide differences between the

WHAT IS PHYSICS OF THE ENVIRONMENT? 7

methods of measurement, the types of pollutants and the five criteria mentioned above. There is clearly no perfect method of measurement.

Varying Sensitivities

Each method has different sensitivities. Consider the water pollutant "Ions" in Table 1.3. The specific ion electrodes mentioned develop an electrical potential (or voltage) when inserted into a solution containing ions. Each type of ion produces a different voltage, so the instrument has differing degrees of sensitivity. There are large variations in the limits of detection in Table 1.4 for each particular ion. When the sensitivity of the specific electrode method for detecting ions in water is described in Table 1.3 as "fair," this is, at best, only a fair approximation. For perchlorate ions, the sensitivity is poor; for cupric (copper) ions, it is excellent.

A method of measurement must have a related goal. This is shown again when trace elements (those occurring at low concentration) in air pollution are measured. Table 1.5 shows the difficulty of selecting a particular method. It indicates the limits of detectability for some trace elements in air pollution, using three types of measurement: neutron activation analysis, emission spectrography and atomic absorption.

Table 1.5 shows average levels of elements found in both an urban (East Chicago, Indiana) and rural environment (Niles, Michigan). As expected, the urban areas are more polluted, but the urban:rural ratio varies with the pollutants. In addition, the methods differ strongly in their limits of detectability.

Many more air trace elements than the seven noted in Table 1.5 exist. As in the case of water pollution, different instrumentation is used for different pollutants and even for different levels of pollutants, such as urban or rural.

The Cost of Detection

There is one factor in instrumentation which to some may be more important than all the others. This factor can govern which type of physics is used to study the environment. The cost of various instruments and analyses can determine their use to a large extent. Table 1.6 brings out the differences. The cost of an analysis of pesticides in water costs over $50. With all other factors being equal, fewer of these tests will be performed than $3-5 conductivity measurements. Everything is not always equal; a large number of expensive mercury analyses have been done in the last few years because of public concern and pressure.

Table 1.3 showed that potential for automation was listed as a criterion for water pollution measurements. The high cost of many of the analyses

Table 1.3 Five Criteria for the Use of Different Types of Measurements for Water Pollutants[4]

Pollutant	Method	Specificity	Sensitivity	Relative Reproducibility	Speed	Automated?
Metals	Atomic Absorption	Excellent	Good	Good	Good	Yes
	Emission Spectroscopy	Good	Good	Good	Fair	Yes
	Polarography	Excellent	Excellent	Good	Poor	No
Ions	Colorimetric	Good	Excellent	Excellent	Excellent	Yes
	Specific Electrodes	Fair	Fair	Fair-Good	Excellent	Yes
Pesticides	Gas Chromatograph (GC)	Fair	Excellent	Poor	Poor	Partial
	GC-Mass Spectroscopy	Excellent	Good	Poor	Poor	No
	Infrared	Excellent	Poor	Fair	Poor	No
Other Organics	Gas Chromatograph (GC)	Fair	Excellent	Poor	Poor	No
	GC-Mass Spectroscopy	Excellent	Good	Poor	Poor	No
	Infrared	Excellent	Poor	Fair	Poor	No
Oil	Gravimetric	Poor	Fair	Fair	Poor	No
	Infrared	Good	Poor	—	Poor	No
	Gas Chromatograph (GC)	Good	Good	—	Poor	No

WHAT IS PHYSICS OF THE ENVIRONMENT? 9

Table 1.4 Detecting Ions with the Specific Ion Electrode[5]

Ion	Lower Limit of Detection (mg/l)
Bromide	0.4
Cadmium	0.01
Calcium	0.4
Chloride	0.4
Cupric	0.006
Cyanide	0.03
Fluoride	0.02
Fluoroborate	0.1
Iodide	0.007
Lead	0.02
Nitrate	0.6
Perchlorate	1.0
Potassium	0.4
Silver	0.01
Sodium	0.02
Sulfide	0.003

Table 1.5 Limits of Detectability of Elements in Air Particulate Matter, and Actual Levels in Urban and Rural Environments[6,7]

Element	Limit of Detectability ($\mu g/m^3$)			Typical Concentration ($\mu g/m^3$)	
	Neutron Activation Analysis	Emission Spectrography	Atomic Absorption	Urban	Rural
Sodium	0.001	0.003	0.001	0.46	0.17
Aluminum	0.001	0.003	0.2	2.18	1.2
Calcium	1.0	0.003	0.002	7.0	1.0
Manganese	0.001	0.011	0.002	0.26	0.06
Copper	0.02	0.01	0.002	1.15	0.27
Nickel	0.01	0.0064	0.006	0.052	0.0
Iron	0.001	0.084	0.01	13.8	1.9

10 PHYSICS OF THE ENVIRONMENT

Table 1.6 Typical Cost of Water Pollution Analyses (as of 1972)[8]

Test	Charge per Sample
Chlorides in Solution	$ 5-7
Solids in Solution	$10-30
Conductivity	$ 3-5
Cyanide	$15-30
Phenols	$15-30
Mercury	$20-30
Other metals	$ 9-25
Pesticides	More than $50
Radioactivity	More than $50

listed in Table 1.6 arises because they have to be done, painstakingly, by hand. If and when some of the processes can be automated, their cost will decrease and more measurements will be performed.

THE NUMBERS OF ENVIRONMENTAL PHYSICS

How are actual calculations made in the physics of the environment? Carbon dioxide in the atmosphere has been produced by living plants since life began, and its concentration changed very slowly until the advent of the Industrial Revolution. At that time, the increased use of machinery and the subsequent spread of the automobile dramatically increased the mass of CO_2 produced. Changes in the concentration of CO_2 in our atmosphere can shift the amount of sunlight passing through it, thus altering the world's climate. Before we can tell whether or not we are controlling or eliminating the excess CO_2 in the air, we either have to measure or estimate it.

Carbon Dioxide and the Biosphere

The mass of CO_2 emitted to the atmosphere by machinery and the internal combustion engine since about 1880 can be described approximately by

$$y_{CO_2} = C_1 \exp \{C_2(t - 1880)\} \tag{1}$$

This is an example of exponential growth. C_1 and C_2 are constants, and t is the year. Records of machine numbers began to be kept around the year 1880.

As an example of calculations, consider the year 1880. When $t = 1880$, then $t - 1880 = 0$. The right-hand side of the equation can be written as $C_1 e^0$. Any quantity to the zeroth power is one. The right-hand side is then equal to C_1, the quantity of carbon dioxide emitted in the year 1880. If C_1 is 10 million kg, for example, the amount emitted for this one year can be calculated. If we also know the value of C_2, the amount emitted for all years can be computed.

What is the significance of the term "exp"? It is a number like any other. The right-hand side of Equation 1 takes a number to a power. The value of "exp" is around 2.718.

We are trying to find out how much the mass of CO_2 has increased in the biosphere. To find the mass of CO_2 discharged into the atmosphere since 1880, we *integrate* Equation 1 with respect to the time t. This involves drawing a curve of y_{CO_2} versus time and finding the area under the curve. The result is

$$A_{CO_2} = (C_1/C_2) \exp\{C_2(t - 1880)\} + C_3 \qquad (2)$$

Here A_{CO_2} is the total mass of CO_2 released since 1880, and C_3 is a constant resulting from the integration.

The constant C_3 can be found by noting that when $t = 1880$, $A_{CO_2} = 0$. In other words, at the beginning of the time period we are considering, the total increase is zero. Substituting this value into Equation 2, we obtain

$$0 = \frac{C_1}{C_2} + C_3$$

Then $C_3 = -(C_1/C_2)$. Substituting this value of C_3 back into Equation 2,

$$A_{CO_2} = (C_1/C_2) \times \exp\left[\{C_2(t - 1880)\} - 1\right] \qquad (3)$$

Not all of this increase in carbon dioxide has stayed in the atmosphere. Some has gone into the biosphere and some has gone into the oceans, which are temporarily excluded from the biosphere for purposes of this discussion. The situation is shown in Figure 1.3. Suppose that the fraction of the total increase remaining in the atmosphere is x, where x is less than 1. Then the actual mass of CO_2 in the atmosphere in any year t is $M_0 + xA_{CO_2}$, where M_0 is the mass of CO_2 in the atmosphere in 1880. There has always been some carbon dioxide in the atmosphere from natural sources, such as plants.

How much atmospheric CO_2 enters the biosphere in a given period of time? A physicist might reason as follows: first, the increase of CO_2 in the biosphere is proportional to the mass in the atmosphere. The more in

12 PHYSICS OF THE ENVIRONMENT

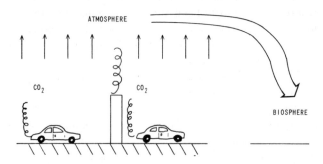

Figure 1.3 Transfer of carbon dioxide between human activities, the atmosphere and the biosphere. A fraction (1 - x) of the increase in atmospheric carbon dioxide goes into the biosphere.

the atmosphere, the more going into plants and animals. Secondly, the increase of CO_2 is also proportional to the time period being considered. For short time periods, a day or a week, twice as much CO_2 will enter the biosphere in two days as in one day.

These points can be written down in words:

(change in mass in biosphere) α (total mass in atmosphere x change of time) (4)

In Equation 4, "α" is a sign representing "proportional to." For example, the number of hours since midnight is proportional to the number of seconds since that time. If the mass in the biosphere in a given year is M, the equation can be written as

$$\Delta M \; \alpha \; (M_0 + xA_{CO_2}) \; (\Delta t) \qquad (5)$$

Here Δ is a short way of writing "change in a quantity." For example, if we represent a tree's height by y, we could write "change in the tree's height since last week" as Δy. The delta symbol Δ doesn't mean a definite change but rather, a relatively small change in the quantity being considered. From the discussion following Equation 3, the total mass of CO_2 in the atmosphere at any time is $M_0 + xA_{CO_2}$.

If quantity C is proportional to quantity D, we can write

$$C = KD \qquad (6)$$

when K is a constant. We can then write Equation 5 as

$$\Delta M = K(M_0 + xA_{CO_2})\Delta t \qquad (7)$$

This is an equation for the change in CO_2 in the biosphere, ΔM. What is the actual mass M for a given year, say 1960 or 1980? The angle of the

WHAT IS PHYSICS OF THE ENVIRONMENT? 13

slope at a point along a curve of M vs t will be related to the ratio of the small change in M to the small change in t. The small changes can be written as dM and dt, where the two terms are not necessarily the same. Then

$$dM = K(M_0 + xA_{CO_2}) \, dt \quad (8)$$

The rules of calculus say that the slope is dM/dt. Once the slope is known, the value of M at any time can be found, again using calculus.

On doing the calculations, the fraction of all CO_2 which has gone into the biosphere since 1880 is

$$Kx \left[\frac{1}{C_2} - \frac{t - 1880}{\exp\{C_2(t - 1880)\} - 1} \right] \quad (9)$$

The constant C_1 disappears from the final result. Reasonable values are 0.30 yr^{-1} for K, 0.35 for x and 0.05 yr^{-1} for C_2. For 1980, t - 1880 is of course 100.

The changes in CO_2 masses in the atmosphere, oceans and biosphere since 1880 are shown graphically in Figure 1.4. The relationships between the three components of the increase are shown. The total mass of CO_2 in all three regions is increasing rapidly.

A physicist must make approximations in order to get an answer to an environmental problem on a world scale. By assuming that the carbon dioxide that humans generate goes into two areas—the biosphere and the atmosphere—a model was devised that produced Equation 9. This equation is reasonably accurate, although there have not been as many measurements made as scientists would like. What is important about the model is that all of the assumptions were written down, and all were converted into mathematics. This does not imply that mathematics is the only key to solving environmental problems. However, its use does force clearer thoughts.

How Accurate is the Measurement?

Most, if not all, the constants used in the discussion of the CO_2 problem were approximations. It made little difference whether K was 0.29, 0.30 or 0.31 yr^{-1}, as long as the final curve was the right shape. For many branches of physics, exact numbers are required. Frequently the numbers are not as exact as desired.

Physics texts occasionally imply that all the data fall on nice straight lines. This is not true in the real world. How is the inevitable error in a typical measurement handled?

Measuring the level of radioactivity in water is an important step in determining environmental quality, especially near nuclear reactors. Table

14 PHYSICS OF THE ENVIRONMENT

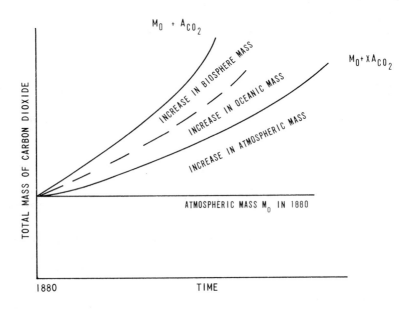

Figure 1.4 Schematic representation of change in the global mass of carbon dioxide since 1880.[9]

1.7 shows the results of an experiment in which two radioactive samples (containing the isotope cesium-137) were sent to 16 different laboratories.

The reported results of samples R and S are shown in the second and third columns of Table 1.7. The measured average values of 61.6 and 49.6 picocuries are close to the known values of 62.5 and 47.7, respectively.

However, there are large variations among the results of the different laboratories. For example, the range for samples R and S is 54.3 and 44.0 picocuries, respectively, where the range spans all 16 laboratories. How is the inevitable error in a typical measurement handled?

Part of the error associated with each of the results is called "random error." If an experiment is done over and over, the results will probably differ slightly from each other. These differences are designated the random error. In a laboratory, systematic errors could be caused by sources such as defective instruments and misuse of instruments. The total error of a given measurement is made up of a combination of the random and systematic error. For example, laboratory E has much higher values for both samples

WHAT IS PHYSICS OF THE ENVIRONMENT? 15

Table 1.7 Calculations of Errors in Measurements of Radioactivity in Water[10]

Laboratory	Results (picocurie per sample)		Calculations					
	Sample R	Sample S	R - S	R + S	R - S - 11.7	R + S - 111.6	(R - S - 11.7)²	(R + S - 111.6)²
A	67.5	54.1	13.4	121.6	1.7	10.0	2.89	100.0
B	39.2	46.6	- 7.4	85.8	- 19.1	- 25.8	364.81	665.64
C	43.2	39.6	3.6	82.8	- 8.1	- 28.8	65.61	829.44
D	62.0	49.6	12.4	111.6	0.7	0	0.49	0
E	93.5	78.4	15.1	171.9	3.4	60.3	11.56	3636.09
F	78.8	65.2	13.6	144.0	1.9	32.4	3.61	1049.76
G	50.7	34.4	16.3	85.1	4.6	- 26.5	21.16	702.25
H	68.8	45.5	23.3	114.3	11.6	2.7	134.56	7.29
I	66.8	51.7	15.1	118.5	3.4	6.9	11.56	47.61
J	61.6	52.6	9.0	114.2	- 2.7	2.6	7.29	6.76
K	64.2	49.5	14.7	113.7	3.0	2.1	9.00	4.41
L	65.4	51.5	13.9	116.9	2.2	5.3	4.84	28.09
M	55.4	45.4	10.0	100.8	- 1.7	- 10.8	2.89	116.64
N	59.9	48.6	11.3	108.5	- 0.4	- 3.1	0.16	9.61
O	49.3	39.0	10.3	88.3	- 1.4	- 23.3	1.96	542.89
P	59.5	47.4	12.1	106.9	0.4	- 4.7	0.16	22.09
Average	61.6	49.9	11.7	111.6				
Total							642.55	7768.57
S_r^2							21.4	
S_r							4.6	
S_t^2								259.0
S_t								16.1

R and S than the true values, and there probably was a large amount of systematic error in its results.

The relevant formulae for the three types of errors are

$$S_t^2 = S_s^2 + S_r^2 \tag{10}$$

where S_t = total error associated with the results
S_s = systematic error
S_r = random error

We cannot calculate the systematic error directly, any more than we can detect a defective instrument without comparing it to a good one. We can, however, compute S_t and S_r by the following formulae:

$$S_t^2 = \frac{(T_1 - T_{av})^2 + (T_2 - T_{av})^2 + ...}{2(n - 1)} \tag{11}$$

where T_1 = total of the paired values R + S for the first lab
T_2 = total for second lab, and so on
T_{av} = average total (111.6 picocuries, at the bottom of column 5 in Table 1.7)
n = number of labs (16)

In the table, T_1 equals 121.6 and T_2 equals 85.8. The numerator in Equation 11 is added for all the labs. Similarly,

$$S_r^2 = \frac{(D_1 - D_{av})^2 + (D_2 - D_{av})^2 + ...}{2(n-1)} \qquad (12)$$

where D_1 = difference of the paired values R - S for the first lab
D_2 = difference for second lab, and so on
D_{av} = average difference (11.7 picocuries, at the bottom of column 4)

In the table, D_1 equals 13.4, and D_2 equals -7.4.

Equations 11 and 12 produce the total error S_t. If S_t is large in comparison to the average values of R and S, this indicates large errors in the measurement. If it is small, it is likely that there were few systematic or random errors encountered.

All data in physics of the environment are not perfect. If only lab E for example, had been used, large errors would have been produced. Its values are considerably higher than average. The fact that there are some errors in almost all data will be shown in the following chapters, because as much as possible the actual experimental numbers will be shown to illustrate a theory.

SUMMARY

A picture has been drawn of some ways in which physicists approach environmental problems. The discussion has been of such topics as the organization of data and results, the approximations necessary to solve a complicated problem such as concentrations of carbon dioxide, and the treatment of errors. Many of these approaches will be implicitly assumed in the following chapters.

WHAT IS PHYSICS OF THE ENVIRONMENT? 17

PROBLEMS

1.1 Construct a table similar to Table 1.1 for common species of wildlife. List different pollutant or environmental conditions which may affect them and the likely effects. Show in which ways the effects could be measured physically.

1.2 Rank the properties listed in Table 1.2 in order of oil value: seawater value. For example, viscosity would rank higher than specific gravity; viscosity at 22°C would rank higher than at 30°C. Disregard properties for which verbal descriptions are given.

1.3 Which of the two types of change in intensity in Figure 1.2 would be easier to read? What does the sensitivity have to do with the slope of the intensity vs concentration curve? Is the sensitivity of the fluorescent method greater than that of the absorption method for the entire range of concentration? Do some sample calculations, assuming that the scales are equal.

1.4 Which of the first four criteria—not counting automation—has the best record, on the basis of Table 1.3? If ions in water were to be measured, which method should be chosen, and why? Which pollutant group, of the five mentioned, would be the easiest, generally speaking, to measure?

1.5 Assume for the sake of argument that each of the seven elements listed in Table 1.5 has equal importance. Which method would be best in terms of the limits of detectability? Which method would be the best for the typical urban environment? The rural environment? Justify the answers.

1.6 A reasonable value of the constant C_2 in Equation 1 is 0.05 yr^{-1}. C_1 is the rate of discharge in 1880. Assume that it is 10^6 kg/yr for simplicity. Then plot Equation 1 as a function of time for about 15 years before and after 1980, on both linear (regular) and "semi-log" graph paper. Explain the difference.

1.7 Plot Equation 3 vs time t beginning in 1880, assuming that C_2 = 0.05 yr^{-1} and C_1 = 10^6 kg/yr. Add up the area under the curve until 1980. The area represents the cumulative mass of CO_2 which has been absorbed by the biosphere since 1880.

1.8 For those familiar with calculus, work through the algebra which led to term 9.

1.9 Using term 9, find the fraction of CO_2 which has gone into the biosphere, from 1880 to 1980. Use C_2 = 0.05 yr^{-1}.

18 PHYSICS OF THE ENVIRONMENT

1.10 Plot $M_0 + A_{CO_2}$ for 10-year intervals from 1880, using Equation 3 and arbitrary values of $M_0 = 10^8$ kg and $C_1 = 10^6$ kg/yr. The curve should resemble the top curve of Figure 1.4.

1.11 A large number of approximations were made to get the value of the fraction in term 9. Name a few. Hints: Is the mass of CO_2 emitted to the atmosphere really increasing exponentially, as implied by Equation 1? Plot car registrations vs time as a check. Is the CO_2 absorptive capacity of the biosphere linearly dependent on the mass in the biosphere? Do all parts of the biosphere absorb CO_2 at the same rate? What would differences in this rate do to the model?

1.12 Plot a "histogram" for samples R and S in Table 1.7, using an interval of 3 picocuries. For example, sample R has 1, 2, 2 and 2 readings falling into the intervals 54-57, 57-60, 60-63, and 63-66 picocuries, respectively. Using this type of diagram aids in the visualization of the values.

1.13 Using Equation 10 and the results for S_r and S_t shown in Table 1.7, find S_s. How large is the systematic error compared to the random error? Could this have been predicted from the lab results alone?

1.14 Calculate S_r, S_t and S_s for the three labs A, B and C.

REFERENCES

1. *Marine Pollution Monitoring Systems,* Space Division, North American Rockwell Corp., no. SD 71-738, (September 1971), p. 9.
2. Klemas, V. "Detecting Oil on Water: A Comparison of Known Techniques," paper no. 71-1068 presented at Joint Conference on Sensing of Environmental Pollutants, Palo Alto, California, November 8, 1971.
3. O'Keefe, A.E. "The New Look in Air Pollution Instrumentation," presented at Analysis Instrumentation Symposium of the Instrument Society of America, Pittsburgh, PA, May 25, 1970.
4. Ballinger, D.G. "Laboratory Methods for the Measurement of Pollutants in Water and Waste Effluents," presented at Joint Conference on Sensing of Environmental Pollutants, Palo Alto, California, November 8, 1971, p. 15.
5. Riseman, J.M. "Specific Ion Electrodes—Versatile New Analytical Tools," *Water Sew. Works* 117(9):IW/12 (1970).
6. Dams, R., J.A. Robbins, K.A. Rahn and J.W. Winchester. "Nondestructive Neutron Activation Analysis of Air Pollution Particulates," *Anal. Chem.* 42(8):865-866 (1970).
7. Zoller, W.H., and G.E. Gordon. "Instrumental Neutron Activation Analysis of Atmospheric Pollutants Using Ge Li Gamma-Ray Detectors," *Anal. Chem.* 42(2):264 (1970).
8. Ballinger, D.G. "Instruments for Water Quality Monitoring," *Environ. Sci. Technol.* 6(2):132 (1972).

9. Munn, R.E., and B. Bolin. "Global Air Pollution—Meteorological Aspects," *Atmos. Environ.* 5(6):385 (1971).
10. Baratta, E.J., and F.E. Knowles, Jr. "Statistical Limits of Accuracy and Precision of Gross Alpha and Beta Radioactivity Measurements in Water," *Radiolog. Health Data Rep.* 12(4):196-197 (1971).

CHAPTER 2

CONSERVATION OF ENERGY, FORCE AND THE ENVIRONMENT

"Eureka!" Archimedes, third century B.C.

"A top does . . . does not cease its rotations, otherwise than it is retarded by the air . . .

If any force generates a motion, a double force will generate double the motion, a triple force triple the motion . . .

If you press a stone with your finger, the finger is also pressed by the stone." Newton, *Principia*

"The conquest of a few axioms (of mechanics) has taken more than 2000 years." Rene Dugas, *A History of Mechanics*[1]

INTRODUCTION

Archimedes' statement commemorates one of the first discoveries of a principle of mechanics—buoyancy. Newton's *Principia,* one of the greatest intellectual accomplishments the world has ever seen, was the summary and synthesis of all the essential principles of mechanics. It had taken two millenia to master these few principles.

Whatever their specialization, most physicists know that their own branch of the subject owes the basis of its measurements to the foundation of physics—mechanics. In mechanics all the forces, energies, pressures and distances which characterize physics are organized.

The importance of mechanics as the cornerstone of physics is recognized in most texts. For example, one of the most widely distributed textbooks devotes almost 40% of its length to it and related subjects. Mechanics would command little space here if we had not been able to illustrate how it works in environmental problems.

The Flywheel

At first glance, mechanics would seem to have no part in solving the problem of air pollution. But a great deal of this pollution is caused by the internal combustion engine. Can the principles of mechanics be used to suggest a substitute?

One proposed replacement is the so-called "super flywheel." A flywheel—a rotating wheel or disc—is a source of kinetic energy, once it has been set spinning. We can draw off this energy, by gears or other methods, just as we can from a battery. There is no pollution from such a flywheel.

The flywheel was used for years in Swiss buses. The experiment was only moderately successful, because of economics and other factors. The laws of mechanics can be used to find the physical factors which can make an improved flywheel. In particular, its stored energy varies with its composition.

The change in kinetic energy is the key to the analysis. The adjective "kinetic" merely refers to motion. The kinetic energy of an object depends on its velocity, but not linearly. If the velocity doubles, the kinetic energy quadruples. In other words, the kinetic energy varies as the square of the velocity.

What about rotating objects? A ball flying through the air has an obvious velocity, but it is not clear if the second hand on a clock has one. However, the hand moves through 360 degrees in a minute. If we call the rate of change of angle the "angular velocity," this concept corresponds to the ordinary linear velocity. The kinetic energy of a rotating body then varies as the square of the angular velocity. For example, if they are exactly the same in every way except their velocity of rotation, the minute hand of a clock will have a kinetic energy $(12)^2 = 144$ times that of the hour hand.

The kinetic energy of an object with an angular velocity ω is defined as

$$E = \frac{1}{2} I \omega^2 \qquad (1)$$

where I is the moment of inertia of the object, a constant.

Figure 2.1 shows different ways a flywheel could be built. The word "flywheel" generally implies a wheel of some sort, but it is simplified into a spinning strip so that the calculations can be done. The centripetal force of the thin rod of Figure 2.1a is

$$F = mr\omega^2 \qquad (2)$$

where m is its mass, and r is its radius.

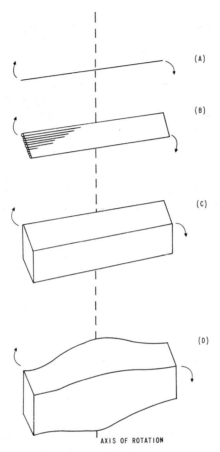

Figure 2.1 Building up a super flywheel from a basic element: (a) basic element; (b) unidirectional strip; (c) stacked strips; (d) optimized shape. Physical quantities to be maximized are calculated from the basic rod in (a), and then they are added together to produce the best shape for conserving energy, shown in (d). Only a thin cross section is shown. The actual flywheel is disc-like in shape.[2]

The force on an accelerating body, according to Newton, is its mass times its acceleration. In turn, the acceleration is the rate of change of its velocity. There is again a difference between straight line and rotating motion. For the former case, the acceleration is just the acceleration. For the latter case, the acceleration is $\omega^2 r$, where ω is the angular velocity defined before, and r is the radius of the circle around which the body is moving.

If ω^2 is substituted from Equation 2 into Equation 1,

$$E = \frac{IF}{2mr} \qquad (3)$$

The rotation creates a force on the flywheel. Since the pressure is the force F divided by area A,

$$F = SA \tag{4}$$

where the pressure per unit cross section of the filament (engineers call it the stress or strength of a material) is denoted by S. A is the cross section of the filament. The strength of materials is easily measured. Equation 4 substituted into 3 produces

$$E = \frac{ISA}{2mr} \tag{5}$$

The energy of the flywheel varies with S, which depends on the construction material. For example, steel has a higher value of S than aluminum or copper.

Which properties of the material other than its shape change E? S does, but I and A do not. The latter two depend only on the rod's thickness, which is not a function of the material being used. Similarly, r is only a length measurement.

But m, the mass of the filament, does change with the material. The familiar equation relating mass, density and volume is

$$m = DV \tag{6}$$

where D is the density of the filament and V is its volume. Equation 5 becomes

$$E = \frac{ISA}{2DVr} = \frac{IA}{2Vr}\frac{S}{D} = k\,(S/D) \tag{7}$$

where the constants in Equation 7 have been lumped into the constant k. Then the energy put into the flywheel varies directly with S/D.

The flywheel can be built up in a number of ways, as shown in Figures 2.1a, 2.1b and 2.1c. The caption notes these are only cross sections. The best shape for storing energy might be expected to be a flat disc. However, the optimized flywheel is in the shape of a center-heavy pancake, as hinted at in Figure 2.1d. But what material should be used to build it?

Table 2.1 shows that although the strengths and densities of materials vary, there is a vast difference between the highest values of S/D (graphite whiskers bounded with epoxy) and the lowest (music wire).

Using the material with the greatest value of S/D allows the most energy in a flywheel. A bus or car can then be built with a longer running times between "recharging." When the recharging time is lengthened, the flywheel vehicle becomes more competitive with those containing an internal combustion engine, thus decreasing air pollution. This sequence shows the connection between physics formulae and reducing pollution.

Table 2.1 Materials for a Super Flywheel[2]

Material	Strength (Newtons/m^2 x 10^8)	Density D (g/cm^3)	Strength/Density (cm^2/sec^2 x 10^9)
Music (Steel) Wire	41	7.95	5.2
Boron/Magnesium	13	2.45	5.2
Steel Wire (bonded with Epoxy)	29	4.97	5.9
Beryllium Wire	17	1.89	9.1
Boron/Epoxy	22	2.08	10.6
Special Glass/Epoxy	23	2.09	10.9
Graphite/Epoxy	20	1.66	11.7
Graphite Whisker/Epoxy	103	1.69	60.8

The engineering problems of the super flywheel (called super because of the advanced materials that will go into it) are formidable. These difficulties include the challenge of building it in a vacuum to reduce air friction, and mounting it on a vehicle. But the discussion shows that even centripetal force can be useful in protecting our environment.

Oil and Mechanics

Physicists often use an "order of magnitude" estimate of quantities; an oil spill is a prime candidate for this type of estimation. To discover how fast the oil spreads with time, consider first the fisheye view of an oil slick, shown in Figure 2.2. In order to compare experiments to the theory to be devised, the volume of the slick must be known.

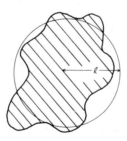

Figure 2.2 Typical shape of an oil spill. Strictly speaking, the blob does not have a radius. However, it is assigned an approximate one (ℓ) so calculations can be performed.

26 PHYSICS OF THE ENVIRONMENT

The first assumption is that the area of the spill fits into a circle of radius about ℓ, as shown in the figure. We do not have to know ℓ exactly for our calculations. The thickness of the spill is almost certain to vary drastically over its area, probably being much greater in the center. An average value is assigned to the oil thickness above the water level. The area of a circle with radius r is πr^2, so the volume V of the slick is the area times the thickness:

$$V = \pi \ell^2 h \qquad (8)$$

The most obvious force on the oil is gravity, which acts equally on all parts. In Figure 2.3, the gravity acts downward, so that the oil sinks into the water until the buoyancy force of the water equals the weight of the oil. The buoyancy force is the upward force the water exerts on the oil. When the downward force of the oil weight equals the upward force, there is no up-and-down motion.

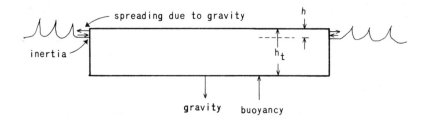

Figure 2.3 Some of the forces acting on an oil spill. Most of the oil is below the surface of the water because of its relatively high specific gravity with respect to water. As the oil spreads out, the thickness of the oil above the water, h, does not decrease as fast as the total thickness h_t. Other forces on the oil, such as viscosity, have not been shown for purposes of simplicity.

The reason for the horizontal spreading is that oil is a fluid. Force applied to a fluid is transmitted equally in all directions. Squash some pancake batter with a kitchen flipper and it will flow outward. So when gravity exerts a force on the oil, it also pushes sideways along the top of the oil, as shown in Figure 2.3.

The physics of oil spreading seems rather simple—the oil spreads out because of gravity. To determine the rate, mathematics is needed.

The vertical forces on the oil are gravitational ($m_1 g$) and buoyant ($m_2 g$). Here m_1 is the total mass of the oil, and m_2 is the mass of water which is displaced by that part of the oil below the water level. This is the definition of buoyant force. Since the oil does not move up or down, the forces must cancel out, or be equal and opposite.

To find m_1 and m_2, note that the mass is the density of a material times its volume,

$$m_1 = DV$$

For oil, density D is D_0. The volume V comes from Equation 8. However, the total thickness of the oil, above and below the water line, is h_t. If this value is substituted into Equation 8, and this changed equation is substituted into $m_1 = DV$, we have

$$m_1 = D_0 \pi \ell^2 h_t \tag{9}$$

The mass of water displaced is

$$m_2 = DV = D_w \pi \ell^2 (h_t - h) \tag{10}$$

where the density of water is D_w. The volume being considered here is that below the water level, and has a thickness of $h_t - h$. Since the two forces are equal and opposite, we have

$$m_1 g = m_2 g$$
$$D_0 \pi \ell h_t = D_w \pi \ell^2 (h_t - h) \tag{11}$$
$$D_0/D_w = (h_t - h)/h_t$$

The equation now gives the relative thickness of oil above and below the water in terms of oil density. It only applies when there is no motion, *i.e.*, the oil and water are in equilibrium. The following discussion considers the case when equilibrium does not exist.

In the horizontal direction, the forces are gravitational and inertial. Although gravity acts vertically, it is transmitted equally in all directions, including horizontal, in a fluid. (Remember the pancake batter.) The horizontal gravitational force is then $m_1 g$.

If gravitation were the only force on the oil, it would spread out over all the oceans instantaneously, since there would be nothing to oppose it. Because this does not happen, there has to be a force opposing this spreading. This is the inertia force, or the resistance objects have to being moved. This force is a reaction rather than the applied force of gravitation. The force F can be described by the equation $F = ma$, where m is the mass being considered and a is its acceleration. To make F independent of the volume of oil spilled, it is converted into force per unit volume, that is, both sides of $F = ma$ are divided by V. Then

$$\frac{F}{V} = (m_1/V)a \tag{12}$$
$$F_I = D_0 a$$

where F_I is the inertial force per unit volume of oil, and other symbols are as previously described. The familiar formula for the mass in terms of density and volume has also been used.

What is the inertial acceleration a? It is not known but there are some clues. Its dimensions are distance/time^2. The horizontal dimension along which the acceleration would take place is ℓ. Let the time equal t. Using this rather rough-and-ready reasoning,

$$F_I = \frac{D_0 \ell}{t^2} \tag{13}$$

This then is the inertial force. Now the gravitational force of the oil above the water must be found. If up is the positive direction, the buoyant force of the water per unit area is $D_w hg$. The quantity h is used instead of h_t because only the mass above the water is being considered. Similarly, the gravitational force on the oil is $-D_0 hg$. The sign is negative because up is defined as positive.

If the oil and water did not move with respect to each other, the two forces would be equal and opposite. Their sum would be zero. Since the oil is spreading over the water, the two forces do not cancel. The total gravitational force is then $(D_w - D_0)gh$. In order to equate the inertial and gravitational forces, their units have to agree. Force per unit area of the gravitational force is changed to force per unit volume. The volume of an object with a pancake shape like the oil slick equals the area times a length. The force per unit area must then be divided by a length.

The gravitational force acts vertically downward, or along the dimension h of the oil. The other dimension of the slick is ℓ, so it is an obvious candidate with which to divide. Since only orders of magnitude are being used, this makes sense. However, it would not be accurate to divide the area exactly by ℓ, because proportionality covers a multitude of approximations. Writing down these mathematical assumptions, the gravitational force per unit volume is

$$F_g \propto \frac{(D_w - D_0)gh}{\ell} \tag{14}$$

$$= \frac{k_1 (D_w - D_0)gh}{\ell}$$

where k_1 is the constant that transforms a proportionality into an equation.

If inertia and gravitation are the only two forces acting on the oil, they can be equated to find out how ℓ varies with time t. There also is a viscous force operating, depending on how slippery the oil is compared to the water, but it will be disregarded for this discussion. Inertia and gravity are the main forces acting when t is small. Equating F_I and F_g from Equations 13 and 14,

$$\frac{k_1 (D_w - D_0)gh}{\ell} = \frac{D_0 \ell}{t^2} \quad (14a)$$

There are now three variables: ℓ, t and h. The final equation should contain only two. Length h can be eliminated. The volume V' of all the oil is $V' = \pi \ell^2 h_t$, where h_t is the total oil thickness. This volume remains constant. Then $h_t = V'/\pi \ell^2$. To write h_t in terms of the unknown h, go to Equation 11. This can be written as

$$\frac{h_t D_0}{D_w} = h_t - h$$

so

$$h = h_t(\frac{D_0}{D_w} + 1)$$

Substituting h_t into this,

$$h = \frac{V'}{\pi \ell^2} (\frac{D_0}{D_w} + 1) \quad (14b)$$

Using this in Equation 14a,

$$k_1 (\frac{D_w - D_0}{\ell}) g (\frac{V'}{\pi \ell^2}) (\frac{D_0 + D_w}{D_w}) = \frac{D_0 \ell}{t^2} \quad (14c)$$

All the terms, with the exception of ℓ and t, are constant. Then

$$\frac{const.}{\ell^3} = \frac{D_0 \ell}{t^2}$$

or

$$\ell^4 = t^2 D_0 \text{ const.}$$

Then

$$\ell = \sqrt{t} \sqrt[4]{D_0 \text{ const.}}$$

If the term under the fourth-root, another constant, is called k,

$$\ell = k\sqrt{t} \quad (15)$$

This shows that the approximate radius of the oil slick, ℓ, varies as the square root of the time. If a slick has a diameter X 100 sec after it starts, it will have a diameter 10 X after 10,000 sec (about 3 hr). This process cannot continue indefinitely, since eventually the slick would cover all the earth's waters. Viscosity forces take over and limit its growth, in the same way that keeps molasses from flowing in January. If these assumptions are correct, for the first interval after the oil is spilled its diameter is governed by Equation 15.

How does this theory compare with experiment? Until the oil spill disasters of the last few years, there was little experimentation on the subject. The data in Figure 2.4 was collected in 1927, and shows how ℓ varies with time for different quantities of oil spilled. If ℓ varies with the square root of t, the results would be a straight line with a slope of one cycle/two cycles, where a cycle is an increase by a factor of 10. A straight line drawn crudely through the points has approximately this slope, although the amount of scattering makes it difficult to draw an accurate assessment.

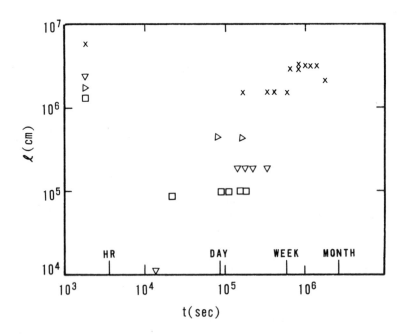

Figure 2.4 Variation of the approximate diameter ℓ of an oil spill with time t. The different symbols represent different spills, in terms of volume of oil.[3]

Oil spills have been discussed from a rather simple viewpoint because all the forces are not known. However, the time variation of ℓ, the approximate radius of the spill, can be estimated. All the other factors which entered into the discussion were grouped into a constant. It did not really matter whether, because of the shape of the spill, the factor π in Equation 8 should have been a 3 or even a 2. This factor, along with others, ends up as part of the constant. These methods of approximation are useful in determining how one variable changes with another—a case physicists are often interested in—but inadequate if an answer correct to the third decimal place is required.

MEASUREMENT

Damping Vibrations

When a workman starts his jack hammer, the irritation to the neighborhood can be intense. Usually when vibration—the repeated deflection of an object over time—is felt, its victims are in a house or building. The standard principle of vibration in physics is demonstrated by an oscillating spring. Its frequency of simple harmonic motion is

$$f = \frac{1}{2\pi} \sqrt{\frac{k}{M}} \tag{16}$$

where k is the "spring constant" or stiffness, and M is the mass attached to the end of the spring. Although a house is not a spring, this equation describes the frequency of vibration of a building to a good approximation. The analogy to k is how easily the building deforms under stress. For example, brick is probably stiffer than wood siding. The mass of the building is analogous to the mass at the end of the spring. These points are discussed further in Chapter 3.

Without going into a mathematical derivation, there are logical reasons for the form of Equation 16. When a spring is stiff, the period of its vibration will be small. Then its frequency, which is the inverse of its period, will be large. This means that f varies in some way with k. Imagine, in turn, an adult and a child on a pogostick. When they jump, the child will vibrate faster than the adult. This means that frequency f varies in some ways with the inverse of M, the mass. The square root relationship of Equation 16 has not been proved, but has been made more believable.

There are means of damping out vibrations, and some are shown in Figure 2.5. The exact mechanisms of these springs and pads will not be detailed, because they will be dependent on the particular case. But typical load vs deflection graphs—it is the building deflections to be

32 PHYSICS OF THE ENVIRONMENT

minimized—are shown on the right-hand side. Each of the four graphs differs significantly in shape from the rest.

Vibrations can be measured easily, just as sound can. But defining vibration pollution becomes just as difficult as defining noise pollution. Different people will react in different ways to varying levels of vibrations. Even in this psychological jungle, human sensitivities to vibration have been measured. Results are shown in Figure 2.6, where the amplitude is the size of the deflection. Physicists are often wary of graphs like this. The terms "annoying," "unpleasant" and so on can hardly be put on a numerical basis.

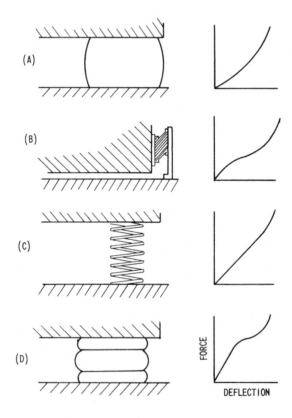

Figure 2.5 Different means of eliminating vibrations. The left-hand side of this figure shows the various means employed, and the right-hand side shows the changes in deflection as the force is increased. The types are: (a) pads in compression, (b) shear, (c) metal spring and (d) air spring.

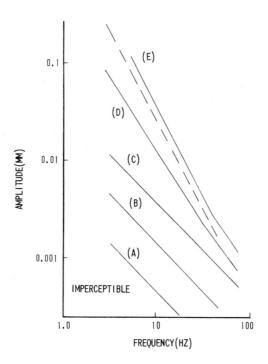

Figure 2.6 Human sensitivities to vibration, as a function of the amplitude (deflection) and frequency. The perceptions are: (a) just perceptible, (b) clearly perceptible, (c) annoying, (d) unpleasant and (e) painful. The dotted line indicates an acceleration of one one-hundredth the acceleration of gravity, or 9.8 cm/sec^2.[4]

Suppose the terms characterizing the different levels of sensitivities are well understood. The degree of human irritation is then dependent not only on the amplitude of the vibration, but also on the frequency. A jack hammer is probably more irritating than a slow tapping.

Although the physical aspects of vibration like amplitude and frequency can be measured, its effect on humans is not as precise. The effect is evaluated by considering its amplitude and frequency together. Vibration is an obvious candidate for the application of mechanics to the environment, since it is essentially small motions of large objects, and motion is an important component of mechanics.

CONTROL

Sludge and Centripetal Force

Sludge is the final result after sewer wastes have been treated and filtered. At that stage, sludge is still mostly water. If most of the water is removed, the final product resembles a solid. It can then be disposed of in landfill sites or, in some cases, used as organic fertilizer.

One way to remove the water is by using a centrifuge. In this method, the sludge is rotated at high angular velocity in a conical bowl. How does this separate out the water? Consider two blobs in the sludge: one mostly solid, the other mostly water. Suppose the first blob has a higher density than the second. If their volumes are equal, the mass of the first will be greater than the second. This may be easily demonstrated using the standard equation for mass, $m = DV$, where D is the density and V is the volume. The equations for the masses of the two blobs are then $m_1 = D_1 V$ and $m_2 = D_2 V$, since the volumes are defined to be equal. If D_1 is bigger than D_2, then m_1 will be bigger than m_2, *i.e.*, the mass of the first blob will be greater.

Suppose the two blobs are dumped into a spinning centrifuge at about the same place. Their angular momenta are then $P = m\omega R$, where m is the mass, ω the angular velocity, and R the distance of each blob from the central axis of the centrifuge.

For each blob in the centrifuge, which is really a miniature merry-go-round, ω is the same, since they are both spinning at the same rate. R is also the same, since they are placed together. The only quantity which would alter their momentum is m, their mass. Since the solid blob has a higher mass, it has a higher momentum. It is flung farther from the central axis of the centrifuge than the water blob. If this experiment is repeated on a large scale, the various blobs would arrange themselves so that those with higher density are on the outside rim of the centrifuge, and those with lower density are nearer the center.

It is then a simple matter of separating water from the sludge by diverting the solids on the outer part of the centrifuge into a bucket. The quantity ω should be as high as possible, since it allows a smaller value of R for the same total momentum. The water and solid can then be separated with a smaller centrifuge.

What is the fraction of sludge solids that can be recovered using a centrifuge, as the centripetal force $F = m\omega^2 R$ varies? This fraction will also vary with the time the sludge spends in the machine. The longer the time, the better the degree of separation of solids and wastes.

Some typical results of experiments in this field are shown in Figure 2.7a and 2.7b. Figure b is really a turned inside out—the centripetal force is used as the x-axis instead of as a parameter accompanying each line. The force is expressed in units of mg, where m is the mass and g is the acceleration due to gravity.

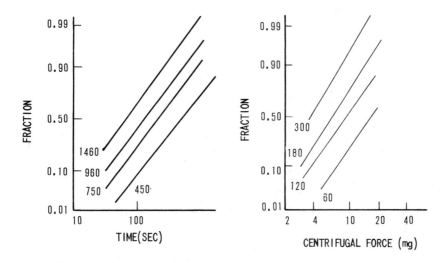

Figure 2.7 (a) Fraction of solids recovered in a centrifuge as a function of time. The numbers attached to each line indicate the centripetal force in units of mg.[5] (b) Fraction of solids recovered in a centrifuge as a function of centripetal force. The units of the latter are in 100 mg. The numbers attached to each line indicate the time in seconds.

The larger the centripetal force, the greater the percentage of solids we can separate out. The same relationship applies to the time of spinning. The graphs can be used to tell the size and types of centrifuges we will need in the future.

Sludge may seem an unlikely physics topic, but as long as our civilization produces wastes which are dumped raw into our rivers, better methods of sludge disposal will be needed. Using the centrifuge looks like a promising method, providing the spinning can be done hard and long enough.

36 PHYSICS OF THE ENVIRONMENT

Cool and Clear Water

There are a number of ways of purifying drinking water. One is called "upflow clarification," and typical apparatus is shown in Figure 2.8. The water to be cleaned passes through the various controls and meters on the right, through a porous plate (to break up the force of the water) and up into the suspension, made up of fine particles which are anything from glass spheres to sand. Dirt in the water tends to cling to the particles, and the water is filtered this way.

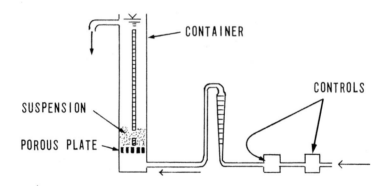

Figure 2.8 The upflow clarifier—a simple method for cleaning water. The suspension must be agitated by the water passing through it, so that the water can be cleaned effectively. The difficulty lies in keeping the water from becoming too agitated. The fluid enters from the right and exits through the container on the left.[6]

How does physics enter into the process? The clarifier is an efficient method of cleaning water, but it has one major drawback. When the velocity of the incoming dirty water is too great, the suspension becomes so agitated that it can overflow the top of the container. On the other hand, the greater the velocity of the water, the faster the process is completed, and the less it costs. The velocity of the water should be as great as possible, consistent with cleaning action. The suspension must also be prevented from becoming too agitated, or having too much "porosity," a measure of how loose the suspension is.

The porosity p can be defined as

$$p = \frac{V_t - V_p}{V_t} \qquad (17)$$

CONSERVATION OF ENERGY, FORCE AND THE ENVIRONMENT

where V_t is the total volume of the suspension for a given water velocity v, and V_p is the total dry volume of the particles of the suspension. When the suspension has a volume 10 times that of its dry condition, its porosity is 0.9. Since volumes can easily be measured in the tube of Figure 2.8, the porosity can easily be calculated.

The porosity has a simple relationship to the velocity. The higher the velocity, the more water entering the suspension. The porosity then should vary directly with the velocity:

$$v = k_1 p \qquad (18)$$

where k_1 is a constant. The equation probably does not hold until the velocity of the water is high enough for the particles to be suspended in the water. There is a cutoff velocity below which the equation is not valid. This situation is shown in Figure 2.9, where each curve has a kink at lower velocities.

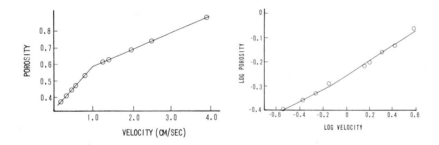

Figure 2.9 (a) Porosity of an upflow clarifier as a function of the water velocity. Note the kink at low velocities. At these velocities the particles are not agitated enough to be really in suspension.[6] (b) Porosity vs velocity, plotted on log-log graph paper. The kink has almost disappeared, due to logarithms being calculated.[6]

Certain calculations, not shown here, show that Equation 18 can be written as

$$v = k_2 p^n \qquad (19)$$

where k_2 and n are constants. Plotting v vs p on log-log paper would then produce a straight line if Equation 19 were correct. If the constant n equals 1, then Equation 19 is of the same form as Equation 18.

If Equation 18 is valid, a straight line on linear (regular) graph paper is obtained. On the basis of Figure 2.9a, it appears that for velocities greater than about 1 cm/sec, Equation 18 wins out. But does it? Figure 2.9b shows that linearity is also found when Equation 19 is plotted on log-log graph paper. This is due to the mathematical properties of this type of graph paper, which tends to smooth out irregularities.

This section is partly concerned with mechanics and partly with how a given theory can be proved experimentally. Physics texts sometimes imply that the subject is a yes-no situation. Problem 2.15 should give an idea of the difficulty in choosing between two alternative theories.

Mechanics, then, has sufficient range to cover everything from flywheels to the spread of oil to sludge and, thus, has a part in numerous environmental concerns.

PROBLEMS

2.1 Plot strength S and density D vs S/D from Table 2.1. Is there any correlation between the three quantities?

2.2 Are there any materials which have greater strength S than those listed on Table 2.1? Compare their S/D ratios to those in this table.

2.3 Would the centripetal force of the super flywheel make the vehicle in which it was mounted rotate? Explain in terms of Newton's law of action and reaction.

2.4 Typical densities of crude oil and seawater are 0.86 and 1.02 g/cm^3, respectively. Calculate what fraction of the oil would be below water level if it were poured on the sea.

2.5 Show that the gravitational force per unit area of the oil above water is

$$(D_w - D_o)gh$$

2.6 What is surface tension? Determine at what point in the spread of the oil it becomes significant. The original paper may be consulted.

2.7 Using Equation 15, plot a theoretical line of ℓ vs t on Figure 2.4. Choose an arbitrary starting point for the line. Does the slope of this theoretical line seem to match the experimental slopes of the data? Why or why not?

2.8 Show that the load versus deflection graph for the metal spring should be a straight line, at least for small deflections. Why does this change for large loads?

2.9 Using physical intuition, explain as much of the detail as possible in the graphs of Figure 2.5. Based on these graphs, if a building were to be isolated from very small vibrations (deflections), which type of support should be chosen? Why?

2.10 Find the equations which describe the amplitudes as a function of frequency for the lines labelled (a) and (b) in Figure 2.6.

2.11 The dotted line in Figure 2.6 denotes a particular acceleration of the building or object being vibrated. Using Equation 16, and any other as required from the mechanics chapter is a physics text, show that the fraction of g, the acceleration due to gravity, that this line represents is about 0.01. Could the concept of acceleration be used to characterize the five levels of sensitivity? Why or why not?

2.12 Using physical intuition, explain why the longer sludge spends in a centrifuge, the greater the degree of separation between solids and water.

2.13 In common usage the acceleration due to gravity is called a "g." What is the greatest number of "g's" a person has experienced? How does this compare with what we do to sludge?

2.14 Suppose a salesman came to the local sewage treatment plant with a centrifuge that would separate out 90% of the solids in sludge in three minutes. The machine had a radius of 3 m and rotated at 600 rpm. On the basis of Figure 2.7 should the machine be purchased, or was the salesman exaggerating?

2.15 Discuss your reasoning in determining whether, on the basis of Figure 2.9, Equation 18 or 19 is valid.

2.16 Suppose that for a particular clarifier, overflow begins when the total volume of the suspension is 20 times that of its dry volume. Assuming that the curve of Figure 2.9a is linear for velocities higher than 1.0 cm/sec, how high can the velocity go before the overflowing begins?

REFERENCES

1. Dugas, R. *A History of Mechanics* (Neuchatel, Switzerland: Edition du Griffon, 1955), p. 11.
2. Rabenhorst, D.W. "Primary Energy Storage and the Super Fly-Wheel," Technical Memorandum TG 1031, Johns Hopkins University, Applied Physics Laboratory, October 1970, pp. 15, 17.
3. Fay, J.A. "The Spread of Oil Slicks on a Calm Sea," in *Oil on the Sea* (New York: Plenum Press, 1969), p. 60.
4. Monk, R.G. "Mechanically Induced Vibration in Buildings," *Environ. Eng.* 51:10,11 (December 1971).
5. Vesilind, P.A. "Estimation of Sludge Centrifuge Performance," *Proceedings of the American Society of Civil Engineers*, Vol. SA3 (June 1970), p. 811.
6. Brown, J.C., and E. La Motta. "Physical Behavior of Flocculent Suspensions in Water," *Proceedings of the American Society of Civil Engineers*, Vol. SA2 (April 1971), pp. 213, 215.

CHAPTER 3

SOUND IN THE ENVIRONMENT

INTRODUCTION

With all the sounds which can be made, it is a wonder that a common denominator can be found to describe them, let alone measure them. While it is an approximation to the truth, such a standard method is used. Before discussing the measurement and control of noise pollution this denominator will be analyzed in some detail, since understanding it is crucial to the whole subject.

The Decibel

Every source of sound radiates power—the units are watts. The range of radiated power from weak sources up to strong sources is so great that a logarithmic scale of units had to be used. This is radically different from the type of units used in other parts of physics. For example, a kilogram is only 1000 grams. Suppose measurements had to be done with grams on one hand and kilo-kilo-kilo-kilograms on the other. One gram would be compared with 10^{12} grams (in scientific notation); a better system of units would be needed.

This is the problem encountered in sound measurements. A typical weak sound, like the hum of a bee, produces about 10^{-11} W, but a strong sound, like that of a nearby jet aircraft, might produce 10^3 W. This is a ratio of $10^3/10^{-11} = 10^{14}$, an overwhelmingly large number. For example, it represents the entire area of the earth compared to one square meter. A scale different from the usual linear gram-kilogram, second-hour relationship is needed.

The problem is solved by the use of the decibel scale. A reference level of power is chosen—call it W_o. If the source of sound has a power W, then the sound power level in decibels is

$$L_w = 10 \log_{10}(W/W_o) \tag{1}$$

How does this equation handle the problem of large numbers? The logarithm of 10^a is defined as a, so each large number is reduced to a more manageable size. One of the main reasons the logarithmic function was used is because of Fechner's law, a rule of human physiology discovered decades ago. It was found that human response to stimuli, such as light and sound, varied approximately as the logarithm of the stimulus. For example, if the physical sound level were doubled, the apparent "loudness" felt by test subjects went up by about 30%, rather than 100%. The logarithm of 2 is 0.301, so the results seemed to follow a logarithmic scale.

The reference level W_o was supposed to be the lowest power that the human ear could hear, and was often taken to be the sound power of the rustle of a leaf. The standard value of 10^{-12} W was chosen to represent it. The sound level of 1 W is then

$$L_w = 10 \log_{10}(1/10^{-12})$$
$$= 10 \times 12 = 120 \text{ dB}.$$

The decibel, usually abbreviated dB, is 1/10 of a bel.

Sound power from ordinary sources is much smaller than the rate of energy production found in other parts of physics. For example, it has been calculated that the total sound power produced by a crowd shouting at a football game—about 50,000 people for 1½ hours—will just barely warm up a cup of coffee.

Moving away from a loudspeaker makes it sound less loud, yet it is still generating the same power. To solve this problem, the concept of *intensity level* is used. A reference intensity level of power per unit area I_o is written, in analogy to W_o, where $I_o = 10^{-12}$ W/m². In analogy to Equation 1, the intensity level is

$$L_I = 10 \log(I/I_o) \tag{2}$$

where I is the sound intensity whose level is being specified in W/m². The intensity level has the same "units" as sound power level. Decibels are not units in the same way that grams or seconds are, but are "units of levels with respect to a reference level."

Consider Figure 3.1 to determine the relation of area to intensity level. An imaginary sphere surrounds a power source of W watts. The sound spreads out uniformly in all directions, and none is absorbed by the air. The second statement is an approximation, since sound is absorbed in air over a distance of a few hundred meters. For relatively short distances, the statement is close to the truth. The area A of the surface of the sphere is

$$A = 4\pi r^2 \quad (3)$$

where r is the radius. All the sound power passes through the surface. The sound intensity I is defined by

$$I = \frac{W}{A} \quad (4)$$

in W/m².

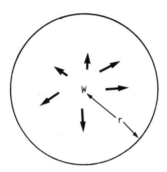

Figure 3.1 An imaginary sphere of radius r around a source of sound power W watts.

On the basis of Equations 3 and 4, the intensity I also drops off as the inverse square of the distance from the source of sound, since substituting Equation 3 into Equation 4 yields

$$I = \frac{W}{4\pi r^2} \quad (5)$$

Limitations of the Decibel

The frequency of the sound being generated has not been discussed. Most sounds have different intensities for different frequencies. This is illustrated for nine sound sources in Figure 3.2. The spectrum human ears can detect is usually taken to be from about 20 Hz to about 20,000 Hz. Typical average sound intensity levels are shown in Figure 3.3. These levels are averages of averages—not only is an average over frequency taken, but an average over time. For example, a home will have a considerably higher sound intensity level at 6 PM than 2 AM.

Loudness, noisiness and other measures of the degree or irritation are essentially psychological and physiological in nature. Different sound level intensities affect people in different ways, as illustrated in Figure 3.4. Some

44 PHYSICS OF THE ENVIRONMENT

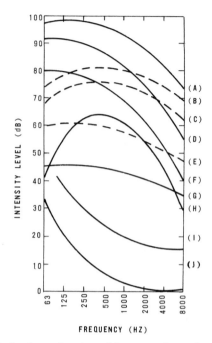

Figure 3.2 Intensity level as a function of frequency for various sound sources. The notations attached to the curves are: (a) jet aircraft at a distance of about 500 m; (b) garbage disposal unit at about 1 m; (c) vacuum cleaner at 1 m; (c) freight train at 30 m; (e) washing machine at 1 m; (f) truck at 6 m; (g) refrigerator at 1 m; (h) human voice at 1.5 m; (i) the desert; (j) approximate human hearing threshold. Because the intensity level falls off with distance, distances for particular sources must be specified.[1]

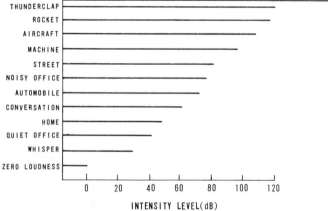

Figure 3.3 Comparative intensity levels of a variety of common sounds, averaged over frequencies.[2]

parts of the population have considerably lower (or higher) thresholds of hearing than others. Changing the sound intensity level will affect some people more than others. At a frequency of 10,000 Hz, there is a range of almost 100 dB in threshold levels between the best 1% and the worst 1% of the population. The graph means, for example, that if one's hearing ability is about average (on the 50% line), it takes a sound intensity of 20 dB at a frequency of 1000 Hz to make a sound noticeable.

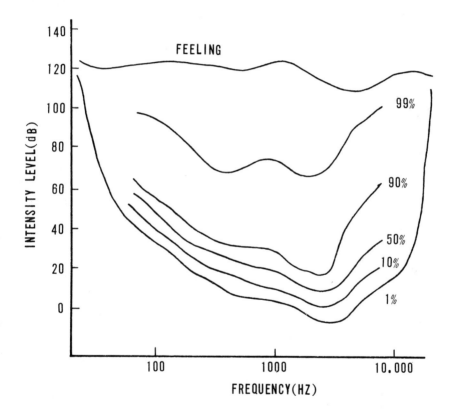

Figure 3.4 Typical threshold hearing levels. These curves mean, for example, that the best 1% of the population in terms of hearing ability requires 20 dB of sound intensity level at 200 Hz to hear sound. Similarly, the best 50% requires 30 dB at 400 Hz to hear, and so on. The 1% curve is often taken as "normal" hearing ability, although this graph shows that this is an exaggeration. The top curve shows the intensity levels at which sounds are felt rather than heard.[2]

46 PHYSICS OF THE ENVIRONMENT

Scales of noisiness and loudness have been devised, but they all depend on psychological and physiological factors, as well as cultural differences. Although watts and decibels are duller words than loudness and noisiness, they can be defined much more clearly because they are based on physical principles.

Adding Intensity Levels

Suppose there are two sources of sound close to each other, each with a sound intensity level, L_I, of 80 dB. What is the total sound intensity level?

Intensity levels do not add linearly, as intensities do. Using Equation 2,

$$\frac{L_I}{10} = \log(I/I_o) \qquad (6)$$

If both sides of the equation are raised to the power 10, recalling that $10^{\log x} = x$, where x is any quantity,

$$10^{(L_I/10)} = I/I_o$$

Then

$$I_o 10^{(L_I/10)} = I$$

The sound intensity for each source is then

$$I = 10^{-12} \frac{\text{watt}}{m^2} \times 10^{(80/10)} = 10^{-4} \frac{\text{watt}}{m^2}$$

The sum of the two intensities is 2×10^{-4} W/m². Substituting this value into Equation 2,

$$L_I = 10 \log\left[\frac{2 \times 10^{-4}}{10^{-12}}\right] = 10 \log(2 \times 10^8) =$$

$$10\left[\log 2 + \log 10^8\right] = 10 \times 8.3 = 83 \text{ dB}$$

Although the intensity doubled, the intensity level went up by only 3 dB.

Highways and Noise

Consider an environmental problem in which the addition of sound intensity levels is important. In Figure 3.5 the upper part represents a highway in which cars are spaced a distance s apart. The first car, at point A, is a distance d from an observer. Each car has an identical sound power of W watts.

SOUND IN THE ENVIRONMENT 47

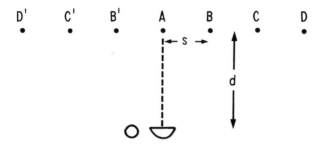

Figure 3.5 Diagram of an imaginary highway, with cars at positions A, B, B′, C, C′, etc., and spaced an equal distance s apart. The observer is at O, a distance d from the nearest car.

For purposes of simplicity, the cars are assumed to be stationary. What is the total sound intensity level at point O of a long chain of cars, at positions B, B′, C, C′, etc.?

For the observer at O measuring the sound intensity from position A, Equation 5 becomes

$$I_A = \frac{W}{4\pi d^2} \tag{7}$$

The power W of each car is the same, so that the total sound intensity at O from cars at positions B and B′ is

$$I_B = \frac{2W}{4\pi(d^2 + s^2)} \tag{8}$$

where the Pythagorean triangle has been used. The factor 2 in the numerator is there because of the two cars. Similarly, for positions C and C′

$$I_C = \frac{2W}{4\pi\{d^2 + (2s)^2\}} = \frac{2W}{4\pi(d^2 + 4s^2)} \tag{9}$$

and so on. Recalling that intensities add linearly, the total intensity of all the sources is

$$I = I_A + I_B + I_C + \ldots \tag{10}$$

$$= \frac{W}{4\pi}\left[\frac{1}{d^2} + \frac{2}{d^2 + s^2} + \frac{2}{d^2 + 4s^2} + \ldots\right] \tag{11}$$

Using the definition of sound intensity level, Equation 2,

$$L_I = 10 \log(I/I_0)$$

48 PHYSICS OF THE ENVIRONMENT

$$= 10 \log \frac{W}{4\pi I_o} \left[\frac{1}{d^2} + \frac{2}{d^2 + s^2} + \cdots \right]$$

$$= \log \frac{W}{4\pi I_o d^2} \left[1 + \frac{2}{1 + (s/d)^2} + \cdots \right]$$

$$L_I = 10 \log \left[1 + \frac{2}{1 + \left(\frac{s}{d}\right)^2} + \frac{2}{1 + \left(\frac{2s}{d}\right)^2} + \cdots \right]$$

$$+ 10 \log \frac{W}{4\pi d^2 I_o} \qquad (12)$$

The second term on the right-hand side of Equation 12 is only the sound intensity level at O from the car at position A. The increase in intensity level is the first large term. Since it is laborious to calculate, this increase is shown in Figure 3.6. As the cars become more separated—or s/d becomes larger—the increase in intensity level becomes smaller. If the distance between cars becomes small, the intensity level increases rapidly. This calculation shows how many sources of sound may be evaluated.

While illustrative of methods used in sound problems, the example is not entirely realistic. Cars differ in the sound power generated. Further, the sound power will depend on such factors as the velocity of the car and whether or not it is accelerating.

MEASUREMENT

Most instruments which measure sound as it affects humans are calibrated to reproduce about the same results as the typical human ear. However, other measurements in physics are made without regard for human capabilities. For example, a light measurement does not take into account the fact that humans see the green part of the spectrum best; an electrical measurement does not note that humans feel shock when receiving a few milliamps of current at low voltage.

Sound Level Meters

If an otherwise perfect instrument is to approximate the ordinary less-than-perfect human ear, its response must be "weighted." For example, at a frequency of 1000 Hz, the instrument would read an intensity level of 80 dB if it were beside a sound source of intensity level 80 dB. At a

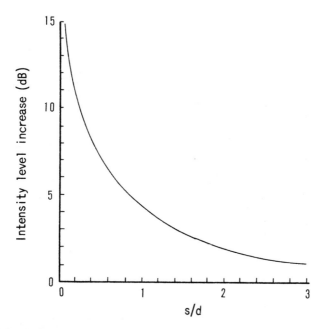

Figure 3.6 Traffic intensity level increase as a function of car spacing s and observation distance d.[1]

source frequency of 200 Hz, though, it would then read only 69 dB. At the latter frequency, the "relative response" would thus be 69 dB - 80 dB = -11 dB. The negative sign indicates a decrease in intensity level. Relative responses in comparison to a perfect measurement are shown in Figure 3.7, as a function of frequency. Two relatively common weighting schemes, labeled A and B, are shown. They correspond to two ways in which the ear perceives the loudness of sounds. The exact weighting is accomplished by a system of electrical filters.

River Velocity

One of the simplest ways of using sound to measure the environment is finding the velocity of rivers by the acoustic velocity meter (AVM). Since water is a precious resource, its use and how much is passing through our rivers should be measured in order to formulate the right conservation policies. Knowledge of river velocity is also important because the rate at which rivers "heal" themselves of pollutant effects depends on this velocity. The principles of the AVM are shown in Figure 3.8. A sound pulse is sent from A and received at C, and another pulse goes from C to A. In

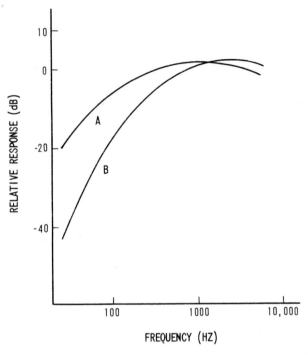

Figure 3.7 Frequency response in standard sound level meters.[1]

the figure, v_p is the average component of the water velocity parallel to the acoustic path. v_L, the average water velocity in the direction of the stream, is required. v_p is found by resolving v_L into components parallel and perpendicular to the line from A to C.

If the standard formula for elapsed time as a function of distance (time equals distance divided by velocity) is used,

$$t_{ac} = \frac{B}{v - v_p} \qquad (13)$$

where t_{ac} = time the sound pulse takes to go from A to C
  v = velocity of sound in water
  B = distance between A and C

There is a negative sign in the denominator because in going from A to C the sound is "fighting its way upstream" *i.e.*, the direction of the velocity v from A to C is opposite to that of v_p. Similarly, the time for the pulse to go from C to A is

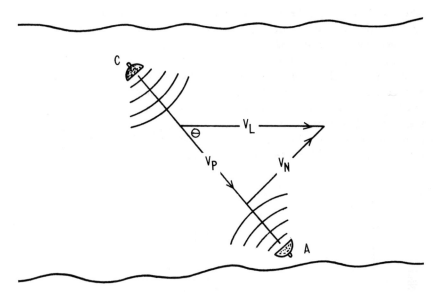

Figure 3.8 The operating principles of the acoustic velocity meter (AVM). C and A are sound pulse receivers and senders. The symbols indicate the components of sound velocity involved.[3]

$$t_{ca} = \frac{B}{v + v_p} \quad (14)$$

The denominator has a positive sign because v_p is now along the same direction as v. As a check on the signs, t_{ac} will be bigger than t_{ca}, because of the stream flow direction. As a result, the denominator of t_{ac} is made smaller with a minus sign. From Equation 13, $v = (B + t_{ac}v_p)/t_{ac}$. Substituting this in Equation 14,

$$t_{ca} = \frac{B}{\frac{B + t_{ac}v_p}{t_{ac}} + v_p} = \frac{B}{\frac{B + 2t_{ac}v_p}{t_{ac}}} = \frac{Bt_{ac}}{B + 2t_{ac}v_p}$$

Then

$$v_p = \frac{B}{2}\left[\frac{1}{t_{ca}} - \frac{1}{t_{ac}}\right]$$

Since $v_p = v_L \cos\theta$ from Figure 3.8,

$$v_L = \frac{B}{2\cos\theta}\left[\frac{1}{t_{ca}} - \frac{1}{t_{ac}}\right] \quad (15)$$

52 PHYSICS OF THE ENVIRONMENT

where θ is the angle between the direction of the stream and the line AC. All that is now required to find v_L is geometry—to find B and θ—and instrumentation—to measure the times t_{ac} and t_{ca} accurately. Both can easily be accomplished.

A series of calibration experiments on the AVM was done in short pipes and Figure 3.9 shows some typical readings. Most of the results are clustered around the zero error mark, indicating that the theory was close to the truth. The calibration is done by measuring the time it takes to fill a bucket of known volume at the end of the pipe, thus getting a flowrate to compare with that found by the acoustic velocity meter.

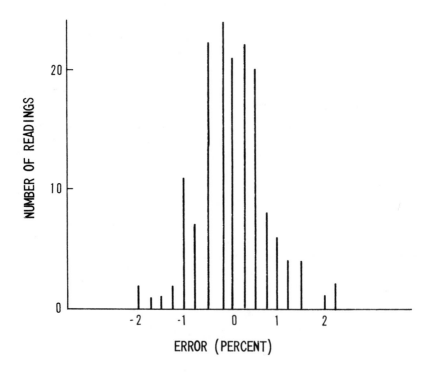

Figure 3.9 Distribution of errors in experiments on the acoustic velocity meter. The plus sign indicates a positive error, and the minus sign a negative error. Very few of the errors are greater than one percent.[4]

The principles outlined here can be used in more than one part of the environment. Knowledge of flow in sewage pipes is critical to the urban environment because so much of our wastes has to be carried in them. The AVM can be employed to measure its rate. By applying it in rivers,

SOUND IN THE ENVIRONMENT 53

information is gained which aids in understanding floods and the measurement of water balance in all parts of the country.

Dust in the Air

One way to measure the mass of dust in the air would be to let the dust settle on a strip of paper, then note the increase in the paper's weight. This would be difficult to do accurately, since it would involve measuring differences in micrograms.

A simpler and more elegant way of measuring the dust accumulation would be to put the strip under tension. In actual experiments, plastic tape is used. The tape acts like a string under tension. A string has a characteristic frequency

$$f = \frac{1}{2\ell} \sqrt{\frac{F}{M}} \qquad (16)$$

where F is the force of tension, ℓ is the length of the string and M is its mass per unit length. Now suppose dust, with mass M* per unit length of the tape, falls on the plastic. The tape's "effective mass" increases, the characteristic frequency drops, and is then

$$f = \frac{1}{2\ell} \sqrt{\frac{F}{M + M^*}} \qquad (17)$$

Consider the strings of a guitar. The thicker the strings (or the greater M is) the lower the frequency of vibration, and the lower the note.

Most dust has a slight electrical charge, and this static electricity enables it to stick to vibrating or vertical surfaces. As a result, the dust stays on the tape.

To find the concentration of dust in the air, all that is required is to find the change in tape vibration frequency as dust settles on it. The frequencies generated fall in the acoustic range 20-20,000 Hz and so can be measured using standard sound instruments. Figure 3.10 shows typical experimental results. The standard method of doing problems in physics, *i.e.*, in a vacuum, seems to give results which do not agree with theory. Theory agrees with experiment when the calculations are modified to take account of the oscillation of the air around the tape. Figure 3.10 can be used to indicate how little dust must fall before the vibration method measures it. For example, changes in frequency can be measured to 0.01 Hz. If curve 1 of Figure 3.10 is used, this change corresponds to about 10 micrograms of dust. The vibration method is then both sensitive, as this last value shows, and accurate, since its principles are simple. Using the physical principles of sound is at least one way of measuring part of air pollution.

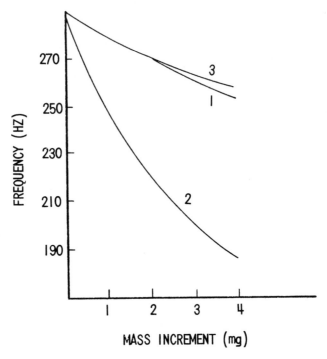

Figure 3.10. Sensitivity of the tape frequency to changes in dust mass falling on it. Curve 1 indicates the experimental results; 2, the theoretical results expected for vacuum; 3, the theoretical results corrected for oscillating air masses.[5]

Bacteria in Sludge

Texts on sound usually deal with vibrations and standing waves. Can any of these concepts be employed on environmental problems?

One use is in increasing the accuracy of bacteria counts in activated sludge from sewage treatment plants. Separating water from sludge was discussed in Chapter 2. In the present example, the sludge is broken down chemically by bacteria. To judge the effectiveness of the treatment, the number of bacteria present must be known. The sludge, however, is usually so thick that counting bacteria is difficult. One way of improving the accuracy of the count is by using a "sonication" technique. In Figure 3.11, a wire is stretched through a container of sludge. Standing sound waves are set up in the wire by an agitator, on the left.

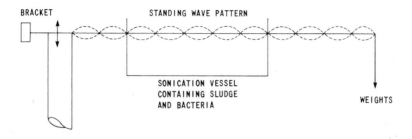

Figure 3.11 Stretched wire sonication apparatus.[6]

By application of these vibrations or standing waves, the particles of sludge and clumps of bacteria were broken up and more bacteria could be counted than before. If the original number of bacteria in a sample of sludge was N_0 and the number after treatment was N, experiments showed that N was generally much greater than N_0. Results are shown in Figure 3.12. The fraction $(N - N_0)/N_0$ increases dramatically with time. The simple use of standing waves can maximize the number of bacteria being counted, though not their actual number, and help to make studies of sewage treatment more accurate.

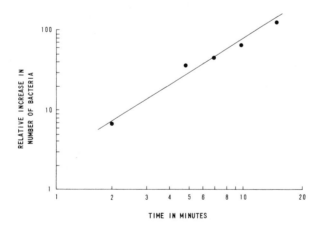

Figure 3.12 Relationship between the relative number of bacteria counted and the time of vibration by the stretched wire method. The relative number is $(N-N_0)/N_0$, where N is the number after treatment, and N_0 is the original number.[6]

Sonic Booms

The "sonic boom" is the sound heard on the ground when an airplane travelling faster than the speed of sound passes overhead. The subject has received much publicity and was a prime reason for the cancellation of the American supersonic transport (SST) by Congress. Other SST's, such as the Anglo-French Concorde, had been prohibited from flying over land by various countries due to the booms they create.

Measurement of sonic booms can be done in terms of dB of sound intensity but because the boom is over in such a short time it becomes difficult to interpret what the measurement means. A simpler way of approaching the subject is to consider it in terms of the change in sound pressure on the ground. Sound waves are simply mechanical waves in the air, so they can be thought of as slight changes (of the order of 0.03% or less) in atmospheric pressure.

The usual form of the pressure exerted by a sonic boom has been described as an "N-wave." An abrupt pressure rise or shock is followed by a fairly linear decline, and then a subsequent pressure rise which restores normal or atmospheric pressure. The N-wave can be called the "wingprint" of the boom, and some typical examples for small, medium and large jet aircraft are shown in Figure 3.13. At first glance the N-waves appear similar, but the differences are visible. Before the boom can be controlled, it must be understood. For example, the maximum pressure change from normal might be the most important variable in a sonic boom. This quantity can be measured from Figure 3.13, and its variation with the type of aircraft specified. Problem 3.24 asks for detailed characterization. Once this has been found, aircraft can be modified so that the effect of sonic booms is reduced or eliminated.

Physicists are often confronted by problems of this type. Just as in other aspects of life, experimental results can be difficult to classify or evaluate.

CONTROL

Walls as Insulators

A truck rumbles down your street. A supersonic jet picks your neighborhood in which to practice dives. How much of this sound enters your home, and how can it be reduced?

Solving the question mathematically is complicated. The problem is simplified by a few gross approximations. Each wall through which sound enters is similar acoustically. In consequence, only one wall will be

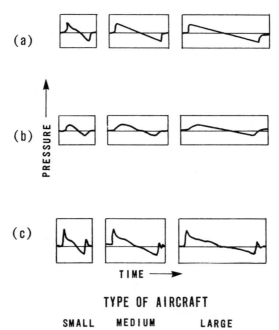

Figure 3.13 Typical pressure-time histories of sonic booms, measured at ground level, produced by small, medium and large aircraft in steady flight. (a) indicates normal N-wave; (b) rounded N-wave; and (c) peaked N-wave.[7]

considered; the roof, other walls and interior of the house will be neglected as the results from one wall can easily be extended to the rest. One more approximation—the wall is assumed to vibrate with the same characteristics as a string. (The vibration of a string was described on page 53). For many purposes, this is not too far from the truth.

Sound can be controlled or reduced in many ways. In particular, how does the mass of a wall affect its sound absorption properties? This is not a completely theoretical question, since the mass of a wall can often be changed easily, but other factors may not be alterable.

The sound intensity is defined as

$$I = k_1 p^2 \qquad (18)$$

where p is the sound pressure and k_1 is a constant. The quantity p is the extra pressure created by the sound waves.

The sound intensity level is, from Equation 2,

$$L_I = 10 \log I/I_0 = 10 \log k_1 p^2 / I_0 \qquad (19)$$

where I_0 is the sound intensity reference level. The difference in intensity levels between outside and inside the wall is then

$$D = 10 \log (k_1 p_i^2/I_0) - 10 \log (k_1 p_o^2/I_0) \qquad (20)$$

$$= 10 \log (p_i^2/p_o^2)$$

where p_i is the pressure inside the wall, and p_o is the pressure outside the wall. Recall that $\log a - \log b = \log a/b$.

The pressure is the force divided by the area. Since the area A of both sides of the wall is the same, Equation 20 can be rewritten as

$$D = 10 \log \frac{F_i^2/A}{F_o^2/A} = 10 \log (F_i^2/F_o^2) \qquad (21)$$

where F_i and F_o are the inner and outer forces, respectively. The outer force F_o is a constant, and depends on the amplitude and frequency of the incoming sound. To find F_i, Equation 16 is written again:

$$f = \frac{1}{2\ell} \sqrt{\frac{F}{M}}$$

Taking this to the fourth power to produce F^2,

$$f^4 = (1/16\ell^4)(F^2/M^2).$$

so

$$F^2 = 16\ell^4 f^4 M^2 \qquad (22)$$

This relationship holds if the wall is analogous to a vibrating string. The quantity ℓ is an "effective length" of the wall, analogous to that of a string, and is a constant. The frequency f will be generally close to the frequency of the incoming sound, another constant. Finally, F in Equation 22 is the inner force F_i. Substituting this result into Equation 21,

$$D = 10 \log (16\ell^4 f^4 M^2/F_o^2)$$

$$= 10 \log k_2 M^2 \qquad (23)$$

where all the constants have been grouped into one constant k_2. M is the mass of the wall. Then

$$D = 10 \log k_2 + 10 \log M^2$$

$$= 20 \log M + \text{constant}, \qquad (24)$$

recalling that $\log x^a = a \log x$. This equation states that the sound reduction should vary with the logarithm of the wall's mass.

Figure 3.14 shows the noise reduction, in dB, for 12 different wall masses per unit area. The noise reduction has been averaged over a wide range of frequencies. Generally speaking, the greater the mass, the greater the sound reduction. The precise mathematical relationship is explored in depth in Problem 3.25. Using this information, we can determine how an increase in wall mass can be compared to changes in muffling ability.

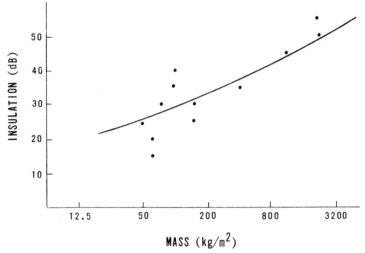

Figure 3.14 Change in insulation value of typical building materials as a function of their mass per unit area. The insulation was averaged over the frequency range 100-3200 Hz.[8]

Openings in Insulators

As a final example, consider a practical case of sound insulation. When a window is opened to let air into a room, the sound intensity level from outside noises increases. Is there any way the differences in insulation can be calculated for different openings in insulators?

Imagine a room with a width of 3 m and a ceiling 3 m high. If there were a solid wall of lead in the center, filling an entire cross section of the room, experiments show that sound would be reduced by about 60 dB in intensity level by the time it reached the other side. This sound reduction is an average over frequencies.

The solid wall of lead is taken out of the room and a cube of lead, 2 m on a side, is brought in. The situation is shown in Figure 3.15, where S is a source of sound, taken to be one frequency for simplicity. It is assumed that the lead block produces no diffracting effects.

The insulated area, presented by the cube, is 4 m², and the uninsulated area is 5 m². Then 55% of the sound passes around the barrier. The

60 PHYSICS OF THE ENVIRONMENT

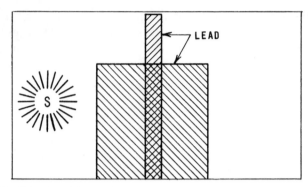

Figure 3.15. Diagram of incomplete sound insulating barriers. Only the longitudinal section is shown. S is the source of sound.[9]

difference in intensity levels, using Equation 2 and assuming that lead is a perfect insulator, is

$$10 \log (I_1/I_0) - 10 \log (I_2/I_0) \tag{25}$$

where I_1 is the sound intensity on the loud side of the lead, and I_2 is the intensity on the quiet side. If it is assumed that I_1 is one unit, then $I_2 = 0.55$. Then the difference, or sound insulation, is

$$10 \log (1/I_0) - 10 \log (0.55/I_0) = 10 \log (1/0.55) \tag{26}$$

$$= 10 \log 1.82 = 2.6 \text{ dB}$$

The experiment will consist of squeezing the lead until it covers a whole cross section of the room, and calculating the change in sound reduction which occurs.

The lead is compressed to a thickness of 1 m. Since its volume is 8 m^3, division shows that the length of a side is now 2.83 m (= $\sqrt{8}$), and its area 8 m^2. The cross sectional area uninsulated is then 9 - 8 = 1 m^2, so that about 12% of the sound passes around the lead. This produces an insulation of close to 9 dB, using the same equations and reasoning as above. If the lead were compressed to a thickness of 0.9 m, the insulation would be close to 19 dB, compared to the theoretical maximum of 60 dB.

The entire problem of insulation gaps can be put on a simpler basis by approaching the problem graphically. The result is Figure 3.16, explained by example. Suppose the average insulation of a wall of brick area 8 m^2, and window area 4 m^2 is to be found. The brick has 45 dB insulation, and the window 25 dB insulation, giving an insulation difference of 20 dB. The answer then is on the line of the graph marked "20 dB." The ratio of the areas of window (lower insulation) to brick (higher insulation) is 1:2. Proceeding to the right of that point of the left-hand axis yields point A. Proceeding down from that point, the loss of insulation is 17 dB. This figure is to be deducted from the brick value of 45 dB, resulting in an average insulation value for the wall containing the window of 28 dB. For the three lead-squashing examples mentioned above, the points on the graph which give the values of sound intensity reduction are points B, C and D.

Figure 3.16 shows that it does not take much of a gap to lose a significant part of the sound insulation. A slightly opened window may do wonders by bringing fresh air into a room, but it also brings in unwanted noise.

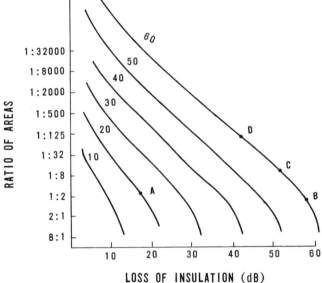

Figure 3.16 The effect on sound insulation of gaps or other areas of lower insulation. The numbers attached to the lines indicate the difference in insulating property, in decibels, of the two areas being considered. The ratio of areas is that of the part of the cross section with lower insulating properties compared to that of higher insulating properties. The loss of insulation is to be deducted from higher insulation. See text for details of calculation.[9]

SUMMARY

Sound is both a physical entity and a potential environmental hazard. If we understand how it is generated, its properties can be used both to measure aspects of the environment and to guard against noise pollution.

62 PHYSICS OF THE ENVIRONMENT

PROBLEMS

3.1 Assuming that the reference level and sound power level are known, find the power W from Equation 1.

3.2 Suppose a person was 5 m away from a loudspeaker radiating 10 W of power. Assuming no absorption from the floor, or sound bouncing off walls, what would be the sound intensity experienced? What would be the sound intensity level?

3.3 If the person moved back another 5 m from the loudspeaker in Problem 3.2, what would the intensity and the intensity level be? Why does the intensity level change so little?

3.4 Consider Figure 3.4 at a frequency of 400 Hz. How much does the sound intensity level have to increase to go from the threshold of hearing of the best 1% of the population in terms of hearing ability, to the lowest 10%, or highest 90%? Plot this range on graph paper for the entire range, 100 to 10,000 Hz. Where is it greatest?

3.5 Since the decibel is really a comparison of levels and not based solely on sound, it could, in theory, be applied to other fields as well. Assume that the height of a typical basketball forward (2 m) is a "reference level." If the distance from the earth to the sun is 150 x 10^6 km, what is its "distance level" compared to the forward's height? Hint: use Equation 2.

3.6 Suppose there were two sources, each with an intensity level of 60 dB. What would be the resulting intensity level of these two sources? Prove algebraically a general statement about two equal sources.

3.7 Find the resulting intensity levels of the combination of three sources with intensity levels of 80, 75 and 70 dB.

3.8 In the example leading to Equation 12, the assumption was that each car radiated sound without any absorption. Suppose half the sound radiated was absorbed into the highway. How would this change Figure 3.6?

3.9 What would the increase in intensity level be, in Equation 12, if s/d became zero?

3.10 Although Figure 3.7 and Figure 3.4 appear to have different y-axes, the two graphs refer in part to the same quantities. Compare line B of Figure 3.7 and the 1% line of Figure 3.4. The relative response of the 1% line corresponds to the difference between what a given group can hear and what it would be able to hear if it had theoretically perfect hearing. For example, at 200 Hz, the 1% group should be able to hear 0 dB but needs an intensity level of 20 dB to hear. Their relative response at this frequency is thus 0 - 28 dB = -20dB. Plot the relative response of the 1% group vs frequency. On the same graph paper, plot the relative response of line B of Figure 3.7. Where is the difference greatest between the two curves? What is this difference in dB? What is the actual ratio of the sound intensities at that frequency? Hint: change dB back into a ratio. Plot the 50% line of Figure 3.4 on the same paper, and answer the same questions. The responses to these queries should indicate how the A and B weighting curves were derived.

3.11 To derive Equation 15, a number of assumptions and approximations had to be made. For example, it was assumed that any small changes in the velocity of the water would not significantly alter the measurements. Name a few other implicit assumptions.

3.12 The ratio of the velocity v_p to the velocity v for an acoustic velocity meter is 1/50. Find the ratio of the time t_{ac} to the time t_{ca}.

3.13 Why aren't all the readings in Figure 3.9 on the zero error mark?

3.14 Calculate the standard deviation for the readings of Figure 3.9.

3.15 Is the average error of all the readings in Figure 3.9 really zero? If not, what is it? Would it be zero if the distribution was a Gaussian curve?

3.16 How fast does the frequency f change as the mass M* changes? This rate is known as the sensitivity. Plot this sensitivity as a function of M*, giving F, ℓ, M arbitrary values of 2 for purposes of the calculation. Start with a value for M* of 2, and continue to M* = 5. For those who know calculus, the sensitivity can be found by differentiation.

3.17 Why do physicists try to formulate experiments such as those whose results are shown in Figure 3.10 only in a vacuum? Why does the theoretical method disagree with the experiment?

3.18 Using Equation 16 and curve 1 of Figure 3.10, with mass increments 0 and 1 mg/cm, find the mass per unit tape length M.

3.19 Is the sensitivity of frequency to changes in mass greater for the "vacuum" or "oscillating air" theory in Figure 3.10?

3.20 Find the velocity of sound in water. Assume that this equals the wave velocity in the sludge. (In reality, the wave velocity will be much lower.) If the container in Figure 3.11 is 5 m long, what frequencies of vibrations will set up standing waves?

3.21 Using Figure 3.12, suppose a typical count of bacteria N_0 before treatment was 10^6 per mℓ. What is the count after 2 minutes of sonication? 10 minutes? Could this increase go on indefinitely?

3.22 What is the probable reason for the apparent increase in the number of bacteria?

3.23 Using Equation 16 for the characteristic or lowest frequency of a standing wave, how can this frequency be increased? Would it increase further the number of bacteria counted? Why or why not?

3.24 Using Figure 3.13, think of a few ways to characterize a sonic boom. There are at least three in common use among physicists. Some of the ways may depend on the shape of the N-wave, and others on the way its areas vary. To prove whether or not the characteristic selected is significant or not, measure it for the 9 graphs of Figure 3.13 and see whether it varies with the size of aircraft and the type of N-wave.

3.25 In comparing Figure 3.14 to Equation 24, does the formula approximate the experimental line? Does the value of the constant enter into the slope?

3.26 In discussing gaps in insulation, it was stated that a cube of lead 2 m on a side would reduce the sound intensity level by 2.6 dB in a room of cross sectional area 9 m^2. Would the results apply to any sound intensity level?

3.27 Derive an expression relating the area of the lead block to its insulating properties. Hint: Because the insulation of a large lead block is 60 dB, for many practical purposes it can be thought of as a perfect insulator.

3.28 How much loss of insulation was there in the lead-squashing experiment if only 1 mm of open space had been left on two sides of the lead? Use Figure 3.16.

REFERENCES

1. Reittinger, M. *Acoustics—Room Design and Noise Control* (New York: Chemical Publishing Co., 1968), pp. 11, 55, 172.
2. Chedd, G. *Sound* (London: Aldus Books, 1970), pp. 16, 17.
3. Smith, W. "Application of an Acoustic Streamflow-Measuring System on the Columbia River at the Dalles, Oregon," *Water Resources Bull.* 7(1):71 (1971).
4. Hastings, C.E. "The LE Flowmeter: An Application to Discharge Measurement," *New England Water Works J.* 84(2):138 (1970).

5. Gast, T. "Acoustical Feedback as Aid in the Determination of Dust Concentrations by Means of an Oscillating Ribbon," *Staub-Reinhaltung. Luft* 30(6):1 (1970).
6. Williams, A.R., C.F. Forster and D.E. Hughes. "Using an Ultrasonic Technique in the Enumeration of Activated Sludge Bacteria," *Effluent Water Treatment J.* 11(2):83, 85 (1971).
7. Crocker, M.J., and R.R. Hudson. "Structural Response to Sonic Booms," *Sound and Vibration* 9(3):455 (1969).
8. Pretlove, A.J. "The Transmission of Outdoor Noises into Buildings," *Environ. Eng.* (March 1970), p. 8.
9. Lord, P., and F.L. Thomas, Eds. *Noise Measurement and Control* (London: Heywood & Co. Ltd, 1963), pp. 167, 168.

CHAPTER 4

HEAT AND THERMODYNAMICS

"With the arrival of your most precious wine, and
with this heat (outdoors), my meditation is
about measuring the aforesaid heat and cooling
the wine. The measurement of the heat is already
reduced almost to perfection, and I have made records
of it for the last 15 days: I shall send a copy
of these by the next post, not having had time to copy
them. I have also found a funnel that quickly cools
wine when it is passed through it. . .
Sagredo to Galileo, July 1613
First systematic measurements
of temperature

"The time has come," the Walrus said,
"To talk of many things:
Of shoes - and ships - and sealing-wax -
Of cabbages - and kings -
And why the sea is boiling hot -
And whether pigs have wings
Lewis Carroll, Through the Looking Glass

INTRODUCTION

The letter from a friend of Galileo is perhaps the first systematic recording of temperature after thermometers were invented at the end of the 16th century. The last quotation listed above alludes to one of the most important environmental problems in which heat is involved today—heated water being discharged from thermal power plants and industries. Lake or river water is taken into the plants and used for cooling purposes and returned to the lake or river at a high temperature, sometimes close to boiling. Even when fish are not killed outright, the hot water promotes the luxuriant growth of algae and other plants, covering the water surface with a green carpet. This often kills life below by cutting off sunlight.

Although thermal pollution is the most obvious example of the interaction of heat with the environment, there are others as well.

Smoke Density

The belching of smokestacks, overfed on coal and oil, used to symbolize prosperity and the good life. Since that time, it has been shown that smoke and dust are literally choking us to death.

There are now laws limiting smoke discharges to legal limits of dust concentration. Unfortunately, these laws are difficult to enforce because of problems in measuring these concentrations.

Many measurements of smoke densities are made by comparing the shade of the smoke to those on a special chart. The chart has varying densities of black lines arranged in different sections. This crude device has been used for decades. Since it cannot be used at night, the time when some companies "blow their stacks" to ensure cleaner plumes during the day, it is clear there has to be a better way.

One method of measuring smoke concentrations is to use some of the properties of heat. Perhaps the simplest is by noting the fact that heat radiation is absorbed, and also scattered, when particles are placed in its way.

On a cloudy day, the particles of water vapor in the cloud absorb some of the heat radiation of the sun and everything becomes cooler. Smoke particles are analogous to the water vapor. If the heat being absorbed by the smoke near a hot surface were measured, an accurate value measure of its density would be obtained.

In Figure 4.1, the left-hand square is a constant temperature source of heat. The source, as well as an absorber, is placed inside a smokestack. The air containing the smoke passes between them. When there is little or no smoke in the air, almost all heat radiated from the hot square passes to the absorber square.

Pure air absorbs comparatively little. When there is a lot of smoke in the air flow, the particles absorb the heat, and the absorber square becomes cooler.

The absorptivity can be defined as the fraction of radiation which is absorbed by the dust particles. It is a pure number, and so has no units. This quantity can be derived from the dimensions of the particles and their density.

Consider the total area of the dust particles, each with an average area A_p. If there are N particles in all, the total area of the dust is

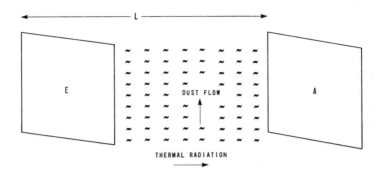

Figure 4.1 Measuring the concentration of smoke and dust by means of heat. The emitter plate, labelled E, is a constant temperature source. The absorber plate, labelled A, is a variable temperature receiver. Both are placed in a smokestack. The thermal radiation from E heats up A. The temperature to which the absorber is heated will depend on the concentration of smoke between the plates.[1]

The area of the heat absorber in Figure 4.1 is A_h, and we define the absorptivity as the ratio of the dust area to the absorber area, or

$$a = NA_p/A_h \qquad (1)$$

To find the quantity N, the mass of the dust particles is written two ways, remembering that the mass equals the density times the volume. The first way is

$$MN = D_p V_p N$$

where M is the mass of each particle, D_p is its density and V_p is its volume. A second way is $V_t D_t$, where V_t is the total volume between the emitter and absorber of Figure 4.1, and D_t is the total mass of the dust divided by the total volume V_t. D_t is then the average density of smoke—the quantity we want to determine.

Equating the two masses,

$$D_p V_p N = V_t D_t \qquad (2)$$

Equating two ways of calculating the same quantity is a major tool of physicists.

The volume of the space between the two plates is $A_h L$, where L is the distance between them, so that

$$N = \frac{A_h D_t L}{D_p V_p} \qquad (3)$$

Substituting this into Equation 1,

$$a = \frac{D_p L A_p}{D_t V_p} \qquad (4)$$

If it is assumed that the smoke particles are roughly spherical in shape, they appear to be circular in cross section. The area A_p is πr^2, where r is the average radius of the particles. Similarly, V_p is $(4/3)\pi r^3$, so Equation 4 can be rewritten as

$$a = \frac{D_p L}{D_t} \frac{\pi r^2}{(4/3)\pi r^3} = \frac{3 D_p L}{4 D_t r} \qquad (5)$$

The absorptivity a is computed from quantities like the dimensions of the air space between the plates L and the average radius of the smoke particles r. The equation may be turned around as

$$D_t = \frac{3 D_p L}{4 a r} \qquad (6)$$

which gives us the average smoke density as a function of the other quantities. Knowledge of quantities like temperature, the way surfaces emit heat, absorptivities, and the way heat is transferred from particles to gases to solids and back again, would tell us how to calculate D_t.

Figure 4.2 shows how the temperature of a particular absorber varies with absorptivity. There is a decrease of the temperature with absorptivity, which is what would be expected physically. When the air is smoky enough, little heat radiation gets through.

Figure 4.3 shows how the absorber temperature varies with dust concentration. As expected, the temperature rises as the dust concentration decreases. The y-axis is logarithmic, so as to bring out the differences at high smoke concentrations. There are large alterations in temperature as the concentration changes, so that this method holds great promise for detecting smoke. An almost century-old method can then be traded for a much more precise one.

Thermal Mapping

Some industries are guilty of thermal pollution—dumping hot water and other liquids into lakes and rivers. How can accurate measurements of temperatures be made at many places near the point of discharge without excessive human labor?

The water temperature at all points on a river may be found by using a property of the heated water. Electromagnetic radiation of differing

HEAT AND THERMODYNAMICS 71

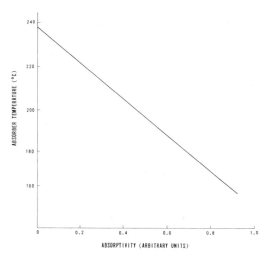

Figure 4.2 Effect of absorptivity on the temperature of the absorber. The more heat the smoke absorbs, the lower the temperature of the absorber. In the model chosen, the relationship is linear. The emitter had a temperature of 550°C.[1]

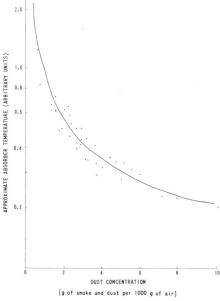

Figure 4.3 Effect of dust concentration on absorber temperature. The inverse of a potentiometer voltage across the two plates in the smokestack has been measured. The voltage varies approximately as the inverse of the temperature. Two inverses cancel when multiplied, producing an approximate measure of absorber temperature. The greater the dust concentration, the lower the absorber temperature.[1]

72 PHYSICS OF THE ENVIRONMENT

wavelengths is radiated from bodies of different temperatures. The energy radiated at a given wavelength by a so-called black body is, by Planck's law,

$$E = \frac{K_1}{\exp(K_2/T) - 1} \tag{7}$$

where k_1 and k_2 are constants, and T is the temperature. As the temperature goes up, the term $\exp(K_2/T)$ goes down, and so does the denominator of Equation 7. As a result, the energy E goes up.

In physics, a black body absorbs all wavelengths of radiation, and by the same token emits all wavelengths. To the observer, it is "black," since all wavelengths are absorbed. To a good approximation, the river also acts like a black body, even though it is not usually black. Equation 7 can then be used. The energy emitted at a particular wavelength is measured to get the temperature T. The actual measurements are made from an aircraft flying over the river. The wavelengths used are in the infrared region, around 10μ in length, or 100,000 Å.

Figure 4.4 shows a small part of data taken from the Wabash River in Indiana. It shows a computer printout of the surface temperatures near an island in the river. The symbols are explained in Table 4.1, with a different symbol used for each range in temperature. For example, if a given part of the river had a temperature of 24.95°C, the symbol printed was Z; if the temperature was 1° higher, the symbol was *. The warmer regions can now be found. The numbers on the left-hand side of Figure 4.4 identify particular lines of computer data.

```
709  • •  * * = | * Z I /  | •       • •    • = — | = = / | — • • •
710  • Z I = | * * / = | • •         •     • — = | = • — | — — —
711  • / X * | * / I M | •      •          • — | * * — | — — —
712     — Z I | I / * / | •                     • | — — = | — • —
```

Figure 4.4 Thermal map of heat radiation in the Wabash River in Indiana. The symbols indicate different temperatures, and are explained in Table 4.1. The blank space in the center is an island, with temperatures greater than 27°C. The areas enclosed by vertical lines had thermometer readings of the temperature taken as a check on remote sensing. The banks of the river and the island tend to be warmer than the center of the stream.[2]

The blank area in the center of Figure 4.4 is an island in the stream. The instrument in the airplane and the computer have no way of telling what is land and what is water, but Table 4.1 notes that if the temperature is above 27.0°C, the space is left blank. Since the land was considerably warmer than the water at the time of the experiment, the island and its banks show up blank.

Table 4.1 Symbols Used to Denote Temperature Ranges for Remote Sensing of Figure 4.4.[2]

Temperature Range (°C)	Symbol
0 to 24.7	M
24.7 to 24.8	$
24.8 to 24.9	X
24.9 to 25.0	Z
25.0 to 25.1	*
25.1 to 25.2	I
25.2 to 25.3	/
25.3 to 25.4	=
25.4 to 25.6	—
25.6 to 27.0	.
27.0 to 33.0	blank

To make sure that the remote sensing system is accurate, tests employing a rowboat were used as a check. The areas where the tests were carried out are enclosed by the vertical lines. The rowboat and airplane results agreed to within a few tenths of a degree, so the remote information was valid.

The area of river covered did not have a thermal power plant beside the stream. However, there are definite regions of cold and warm water. Since these regions can be differentiated, the same technique may be used to check on thermal polluters.

Remote sensing, employing a wide variety of electromagnetic and optical methods, is one of the most important techniques for measuring the condition of the environment. Everything from earth satellites to one-engined aircraft can carry sensing instruments. Since thermal pollution is one of the more difficult aspects of wastes to evaluate properly, remote sensing is a good candidate for helping to measure and combat it.

MEASUREMENT

Thermal Accounting

Is thermal pollution just a matter of plumes of hot water trailing off into lakes and streams, or can the entire receiving body of water have its temperature raised? It depends on the size of the lake or river. Suppose the question is whether a body of water will rise in temperature by a thousandth of a degree Celsius—which means that almost nothing will happen to the life in it—or 5°C, which could mean the destruction of all its life. Such quantities as its size, the ability of water to absorb heat, the heat which is being put in, and that which is being removed must be known.

Consider a typical set of measurements conducted on Lake St. Croix, between Minnesota and Wisconsin, as shown in Figure 4.5. Before quantities like the temperature increase can be calculated, all the ways in which heat can enter or exit from the lake must be tabulated. The compilation is sometimes called the heat budget. This type of budget is always balanced; all the heat that goes in must equal all that comes out.

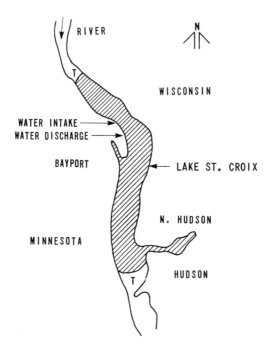

Figure 4.5 Map of the Lake St. Croix region between Wisconsin and Minnesota. The lake is fairly small—about 8 km in length—so the effect of thermal discharges can be relatively great. The letter T indicates a water temperature recorder. The "lake" is a natural widening of the St. Croix River.[3]

There are three major ways heat enters or leaves the lake:

    a. heat moving across the surface of the water
    b. heat entering or leaving via the river which feeds the lake
    c. heat being added by the thermal power plant

There are four parts to the first category. First is radiation. Any body whose temperature is above that coldest of cold temperatures, absolute zero, radiates some heat to its surroundings. Even ice on the lake in the frozen depths of winter is radiating. The process works both ways—radiation from the sun warms the water. Second is convection, or the motion of the air above the water due to the heat it absorbs. Third is the evaporation of the water. Evaporation occurs even when water is not close to boiling. The water molecules which leave the surface tend to be those which move faster. As a result, heat is transferred. Fourth, conduction occurs when heat energy is transferred from one molecule to the next by collision. The water in the lake and the air above are in contact. Since the air can be warmer or cooler than the water, there can be a gain or loss of heat by this process. Categories b and c consist of only one component each.

How are the half dozen or so components of the heat budget calculated? Many of the equations are based on the standard formula relating heat energy and change of temperature:

$$\Delta Q = mc\Delta T \qquad (8)$$

where ΔQ = heat absorbed or emitted
 m = mass of the body being warmed or cooled
 c = specific heat or the heat absorbed or given off per unit mass per unit change of temperature
 ΔT = change of temperature

The symbol Δ signifies a change in the quantity following it. If the hot water issuing from the thermal plant is being considered, ΔT is the change in temperature between the cool water it takes in and the hot water it puts out, and m is the mass of water warmed.

Equation 8 is representative of about six simultaneous equations describing the same number of heat transfer processes, such as radiation and convection, at the same time. For example, Equation 8 relates the heat output of the thermal plant to the temperature change. As more and more heat is put into the lake, the temperature of the cool water entering the plant will rise. This will change ΔT and thus ΔQ. Meanwhile, since the whole lake has become warmer, more heat will radiate from its surface and the radiation will also change. All the equations of heat change when one equation changes.

The most obvious physical quantity is the surface temperature of the water, as shown in Figure 4.6. Only the summer is considered. Although winter temperatures are important, there is less of an effect from thermal pollution at that time of year, for obvious reasons.

In the region being considered, the temperature gradually moves up from about 18°C in June to about 26°C in July and August. It decreases to 4°C by the end of October. The summer heat of course plays a large part in the increase.

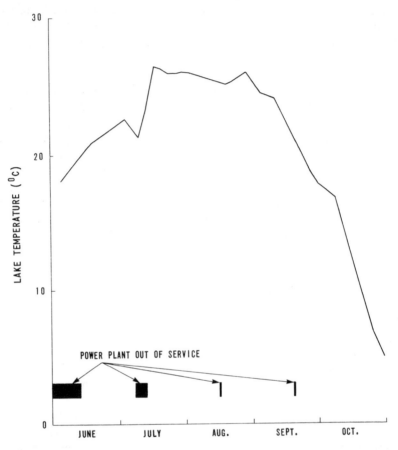

Figure 4.6 Average surface water temperature in Lake St. Croix in the summer and and fall. Although the surface temperature seems to rise when the thermal plant is out of service in June, it is probably due to an increase in outdoor temperature with the onset of summer.[3]

How much of an effect does the thermal plant have? The times when the plant was turned off (most likely due to mechanical problems and not environmental concern) are given at the bottom of Figure 4.6. Problem 4.18 asks for an evalution of this effect on lake temperature.

Figure 4.7 shows cumulative heat put into the lake by the plant. To determine the heat added to the lake each month, subtract the previous month's total. In June, about 0.6×10^{15} J were added; in July about $(1.1 - 0.6) = 0.5 \times 10^{15}$ J and so on.

HEAT AND THERMODYNAMICS 77

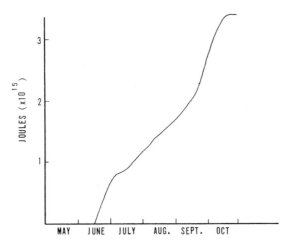

Figure 4.7 Cumulative heat put into Lake St. Croix by a thermal plant. July's value will always be greater than or equal to June's, and August's will always be greater than or equal to July's. There is a fairly steady rise in the accumulated heat, except for a few days in July and at the end of October.[3]

An example of the changing heat transfers is illustrated in Figure 4.8, which shows the heat inputs of the river flowing into the lake. From June until mid-July, the river is warmer than the lake, so heat is being added; after mid-July, the river is colder than the lake, so heat is subtracted.

Heat transfers from other natural causes, such as radiation from the sun, evaporation and conduction have not been shown. Figure 4.9 shows how all heat transfers into and out of the lake change with time. The effects of the power plant are subtracted to calculate the bottom curve. The maximum heat accumulation is in the second half of July. It drops off then, and by the middle of September the amount of heat put into the lake since June almost equals the amount taken out.

The quantity of interest is that actual lake surface temperature and what it would be if the power plant were removed. Figure 4.10 shows both theoretical and experimental situations. The temperature for no power plant, one plant and two plants increases relatively smoothly, with a kink around July 9, to a maximum around the beginning of August, and declines until the end of September. Obviously, the theoretical surface temperature is highest with two power plants, and generally lowest with none. The way in which the temperatures vary is irregular. Problem 4.22

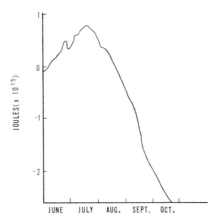

Figure 4.8 Cumulative heat taken out or put in by the flow of the river into and out of the lake. Figure 4.5 shows geographical details. Because the river may be warmer or cooler than the lake, heat can be added or subtracted. The curve represents a total effect over the summer. More heat is added than subtracted until about the middle of July, although the cumulative effect does not show up until the middle of August. By the middle of September, the amount of heat subtracted by the river is within about 0.8×10^{15} J of that added by the thermal plant.[3]

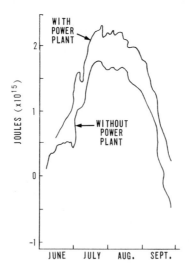

Figure 4.9 Heat accumulation in Lake St. Croix with and without thermal plant heat input. The bottom curve takes account of all heat inputs and outputs, except that of the thermal plant. The top curve takes the plant into account. The heat from the plant averages about 0.5×10^{15} J in August. By mid-September, the accumulated heat without the plant is the same as it was in early June.[3]

explores some conclusions which can be drawn from Figure 4.10. The lake surface temperature can be as much as 4°C above that expected for no thermal plant heat input.

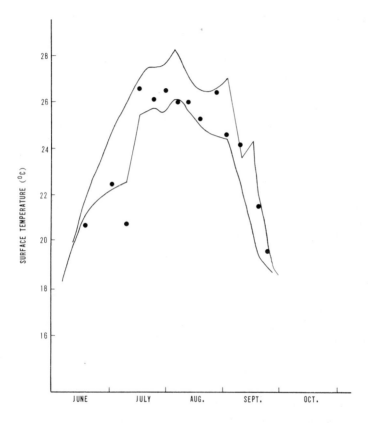

Figure 4.10 Surface water temperatures of Lake St. Croix, both theoretical and measured. The top curve denotes the theoretical temperatures if there had been two thermal power plants on its shores, and the bottom curve temperatures with no power plants. The points indicate the measured temperatures for one plant.[3]

In this section, some of the ways in which heat is added to and subtracted from a lake have been explored. A power plant can add heat which is comparable in magnitude to some natural processes. Because of the multitude of these processes, calculating exactly how much a given power plant will warm up a lake is difficult. The calculations for Lake St. Croix show that the surface temperature could drop by over 1°C over most of the summer if the power plant were removed. However, the temperature near the plant water outlet has a much greater increase than 1°C, harming plant and fish life.

Power Plants and Weather

If enough heat is put into lakes and rivers, can it change the weather as well? The rate of evaporation from water, which produces rain, depends on the temperature. The temperature of water in turn depends on the heat put into it. Almost all the heat normally comes from the sun. Is a balance being upset when power plants warm up lakes? To answer this, the relative magnitudes of temperature increases and heat transfers must be known.

In the last section, Lake St. Croix was a relatively small lake, so power plant effects could easily be seen. In a larger lake, like Lake Michigan, the effects of heat inputs will tend to be smaller.

Some brief calculations of how the weather will change can be made. Assume that all the thermal plant heat ejected into the lake is mixed thoroughly. Rearranging Equation 8,

$$\Delta T = \frac{\Delta Q}{mc} \qquad (9)$$

Reference books show that m, the mass of Lake Michigan, is about 4.7×10^{18} g. The specific heat c of water is 1 cal/g/°C. Suppose the power of a typical power plant is about 1000 MWe, or 1000×10^6 W = 10^9 W. Since one watt times one second is defined as one joule (J) of energy, each hour $3600 \times 10^9 = 3.6 \times 10^{12}$ J are produced. Proceeding with the multiplication, the power output per year is 31×10^{15} J.

How much heat is generated from this power output? A rule of thumb is that power plants have an efficiency of about one-third. For every three units of energy used, one leaves as electrical power, and two leave as waste heat. The waste heat is then twice the power output, or 62×10^{15} J of heat.

When heat energy is dealt with, it is customary to use calories, where 1 cal = 4.18 J. The waste heat output of the thermal plant is then

$$62 \times 10^{15} \text{ J} \times 1 \text{ cal}/4.18 \text{ J} = 15 \times 10^{15} \text{ cal}$$

This value for ΔQ, the heat into the lake, is substituted into Equation 9, getting

$$\Delta T = \frac{15 \times 10^{15} \text{ cal}}{1 \text{ cal/g/°C} \times 4.7 \times 10^{18} \text{g}} = 0.0032°C \qquad (10)$$

The total heat rise in the lake from this power plant will be about three thousandths of a degree, which is hardly measurable. This rough calculation gives an idea of the magnitudes involved.

It has been assumed that the added heat was mixed uniformly throughout the entire mass of the lake. Most of the heat tends to stay in the top 20 m or so (the epilimnion) of the lake. Water is a relatively poor conductor of heat, and so the heat on top often does not get to the bottom.

Consider only the top layer of water. In Equation 10, the mass of the top layer is

$$m = DV = DAd \qquad (11)$$

where D = density of water
         V = epilimnion volume
         A = lake area
         d = epilimnion depth

The lake area is about 6×10^{14} cm^2, so

$$m = 1 \text{ g/cm}^3 \times 6 \times 10^{14} \text{ cm}^2 \times 2{,}000 \text{ cm} = 1.2 \times 10^{18} \text{g} \qquad (12)$$

Substituting this value of m into Equation 10,

$$\Delta T = \frac{15 \times 10^{15} \text{ cal}}{1 \text{ cal/g/}^\circ\text{C} \times 1.2 \times 10^{18} \text{ g}} = 0.0125 \,^\circ\text{C}$$

The temperature increase in the lake surface layer is about four times that of the entire lake, but is still a very small quantity. It is probably too small to affect the weather significantly.

Can the heat input evaporate enough water to cause rain or change the humidity around the lake? Suppose all the heat goes into evaporating water. The heat required to evaporate a mass m of water is

$$\Delta Q = mL \qquad (13)$$

where L is the latent heat per unit mass. The latent heat is required to change water from the liquid state to a gas—water vapor. The unknown in Equation 13 is m, the mass of water is evaporated each year. Then

$$m = \frac{\Delta Q}{L} \qquad (14)$$

What is the actual depth d of water which is evaporated? Rewriting Equation 11,

$$m = DAd \qquad (15)$$

where D is the density of water, A is the area of the lake, and d is now the depth of evaporated water. Substituting this into Equation 14,

$$DAd = \frac{\Delta Q}{L}$$

$$d = \frac{\Delta Q}{ADL} \tag{16}$$

For a 1,000 MWe thermal plant, $\Delta Q = 15 \times 10^{15}$ cal. The area A of Lake Michigan is 6×10^{14} cm^2, the density D is 1 g/cm^3, and the latent heat L is 540 cal/g. Substituting these values in Equation 16,

$$d = \frac{15 \times 10^{15} \text{ cal}}{540 \text{ cal/g} \times 6 \times 10^{14} \text{ cm}^2 \times 1 \text{ g/cm}^3} = 0.046 \text{ cm}$$

The depth of Lake Michigan that evaporates as a result of this thermal plant heat input is 0.046 cm per year. How does this compare with the natural evaporation rate? The level of the lake remains about the same from year to year, except during flooding. If rivers flowing into and out of the lake are disregarded, about 76 cm/year (30 in./yr) of rain falls on it. Since the level remains about the same, about 76 cm/year must disappear by evaporation. So the ratio of thermal plant evaporation to natural evaporation is about

$$0.046/76 = 0.06\%$$

again a small proportion. It is then unlikely that much water will be evaporated due to manmade causes from only one plant. There are, of course, many electricity plants around this lake. The amount of rainfall and snowfall depend to some extent on the amount of water evaporated, so there probably will not be any large change in them as a result of the heat from this single plant.

On a large scale at least, thermal plants do not seem to be able to affect weather or the general conditions of big lakes. Most instances of thermal pollution occur on a smaller scale than the 22,000 square miles of Lake Michigan.

SUMMARY

The heat balances in nature are often delicate. By using the relationships of thermal physics, a better understanding of these balances can be obtained.

HEAT AND THERMODYNAMICS 83

PROBLEMS

4.1 In the example discussing the measurement of smoke concentration, would the process be affected by conducting the experiment at night?

4.2 Explain in simple physical terms the derivation of Equation 1. Why is it assumed that all the heat radiation which strikes a particle is absorbed? What has been assumed about the absorptivity of air?

4.3 Using an encyclopedia if necessary, find the densities D_p of typical smokes and dusts. If this information is unavailable, use as a substitute the density of soil. Is it correct to disregard the density of air in Equations 1-6?

4.4 Based on Equation 5, if the mass of particles per unit time passing through our instrument doubled, what would happen to the absorptivity?

4.5 Would changing the distance between absorber and emitter allow us to measure smaller smoke densities? Explain.

4.6 In Figure 4.2, why is the temperature of the absorber not equal to that of the emitter (550°C) when the absorptivity is zero?

4.7 In order to use the absorptivity method of detecting smoke, the difference between dust concentrations must be found. Suppose a difference of 0.05 units on the y-axis of Figure 4.3 was required before two concentrations could be differentiated. What would be the maximum concentrations that could be told apart using this criterion? As an example, the y-axis units for smoke concentrations 9 and 10 parts per thousand are about 0.21 and 0.23. Their difference is thus 0.02, so by using the above criterion these two concentrations could not be told apart.

4.8 Investigate the effect of excess dust particles in the atmosphere and the temperature of the earth. As a start, look up an encyclopedia article on Krakatoa, the volcano which exploded around the turn of the century. What relation does this have to air pollution?

4.9 Using Equation 7, find temperature T in terms of energy E. The constants K_1 and K_2 are related to the energy wavelength. K_1 varies as the inverse fifth power of wavelength, and K_2 as the inverse first power.

4.10 If the thermal pollution problem were compounded by an oil film problem, would measurements by the radiation method be affected?

4.11 Simplify the temperature distribution of Figure 4.4 by dividing it into just two groups: below 25.3°C (the first seven symbols of Table

4.1), and above 25.3°C (the other four symbols). On Figure 4.4, pencil in those sections of the water having temperatures below 25.3°C. Is the water on one side of the island cooler than on the other? Why or why not?

4.12 The "temperature snapshot" of Figure 4.4 was taken in September. Would there be a better time of year to do the experiment if there were a thermal polluter on this stretch of river? Explain.

4.13 Suppose a thermal polluter decided to pipe hot water some distance into the river, rather than discharge it at the bank, and have the pipe submerged. Would this form of pollution remain undetectable with the remote infrared detection method? Only surface temperatures can be measured with this system.

4.14 What ways other than those mentioned in the text could heat—or energy—be put into or taken out of the water? For example, do living things produce heat?

4.15 Describe what the terms in Equation 8 would mean for the river flowing into Lake St. Croix. Refer to Figure 4.5.

4.16 To keep coffee as hot as possible ten minutes after pouring it, should cream be added immediately, or after the ten minutes have passed?

4.17 Plot the temperature changes in Figure 4.6 from mid-month to mid-month. Which change is greatest? Why does it occur at that particular time?

4.18 Is there any effect in lake temperature when the thermal plant is not working? Why is this true?

4.19 One way of determining how much heat is being put into the lake each day is to find the slope of the curve at each point of Figure 4.7. For June the rate of heat input was about 1.3×10^{15} J/month. Find the rate of heat input at the midway point of each of the following months. Which month has the greatest rate? What is the rate in late October?

4.20 Find the slopes and thus the rate of heat addition or subtraction for the middle of the months shown in Figure 4.8. How do these rates compare with those for the power plant, shown in Figure 4.7? What happens to the heat transfer around July 17?

4.21 Plot the difference of the two curves in Figure 4.9, for heat accumulation with or without the power plant, as a function of time. For simplicity consider only the 1st and 15th of each month. When is the difference largest? Why does it occur at this time of year?

Why are the differences smaller than expected on the basis of Figure 4.7?

4.22 Plot the difference in temperatures in Figure 4.10 between (a) two plants and one; (b) two plants and none; and (c) one plant and none. At what time of year are the differences greatest? Investigate the biological effects that the difference in temperatures in case (a) could have. Are they significant? Are there significant differences between the measured temperatures for one power plant and the theoretical temperatures? Why do these differences exist?

4.23 Assume that half of all electrical power came from plants that were one-quarter efficient, and the other half from plants that were one-third efficient. What is the proportion of waste heat?

REFERENCES

1. Reisbig, R.L. "Detection and Analysis of Particulate Pollution," in *Proceedings of the Centennial Symposium on Industrial and Urban Wastes,* University of Missouri at Rolla, Rolla, Missouri (February 1971) Vol. 1, pp. 63, 67, 69.
2. Atwell, B.H., R.B. MacDonald and L.A. Bartolucci. "Thermal Mapping of Streams from Airborne Radiometric Scanning," *Water Resources Bull.* 7(2):238 (1971).
3. Stefan, H., C.S. Chu and H. Wing. "Impact of Cooling Water on Lake Temperatures," *Proceedings of the American Society of Civil Engineers, Journal of the Power Division.* 98(PO 2): 255, 258, 259, 269, 270 (1972).

CHAPTER 5

ELECTRECOLOGY: ELECTRICITY AND MAGNETISM

"In 1775...Cavendish had compared the conductivities of various substances by the simple method of estimating the intensity of the electric shock which he received."
R. A. R. Tricker, *Early Electrodynamics*, Pergamon Press, Oxford, 1965, p. 7

INTRODUCTION

Since the time of Cavendish, one of the founders of modern electrical theory, our society has become dependent on electricity. Can this pervasive influence be translated into solving environmental problems?

Because electrical measurements are one of the simplest ways to describe physical quantities, this chapter contains many examples. These measurements are used to deal with everything from how fish react to pollutants to finding oil slicks.

Fish and Electricity

Fish can be called a "leading indicator" of water pollution, accumulating pollutants absorbed at the lowest end of the food chain. Concentration of pollutants in fish tends to be much higher than in the surrounding water. For example, the first cases of mercury pollution were found after people ate fish contaminated with the metal, and not from measurements of mercury in the water in which the fish swam.

The experiment to detect pollutants in fish is carried out in tanks through which an electrical current, small enough not to harm the fish, is passed. When it moves all or part of its body, the potential difference between the two ends of the tank, to which electrodes are attached, is changed slightly. As the water calms down, the potential returns to its normal value, but the changes can be measured.

This should be made clearer by Figure 5.1, where potential in a tank containing a bluegill changes with time. The spikes in the chart are due to breathing or opercular movements. The midway change in the direction indicates the fish turned around.

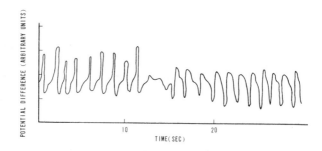

Figure 5.1 Change of potential difference between electrodes in a tank containing a bluegill fish. The regular spikes indicate breathing or opercular movements. The fish turned around in the tank at the 12-second point.[1]

Suppose the breathing rate of the fish were correlated with the pollutant concentration. This possible correlation is shown in Figure 5.2. Fish C and D were in tanks with a concentration of 0.5 mg of copper per liter of water. For Fish A and B, the control group, the breathing rate remained constant. However, for C and D it increased. Curve C ends after two days because the fish died. The difference between the two sets of curves is dramatic.

A simple electrical measurement in a fish tank can indicate how fish react to polluted water. If different species were taken from rivers it might be possible to use their rate of breathing as a guide to the degree of pollution in the water.

Water Conductivity and Pollution

A more direct way of using electricity to measure water pollution is by determining water conductivity. The units of resistivity are ohm-cm. The conductivity is the reciprocal of the resistivity, so its units are reciprocal. A $(cm)^{-1}$ is $1/cm$, but $(ohm)^{-1}$ is mho.

Pure distilled water has a conductivity of about 0.055 μmho/cm; natural water has a conductivity of from 100-200 μmho/cm, because of impurities. For the case under consideration, its actual value was 180 μmho/cm.

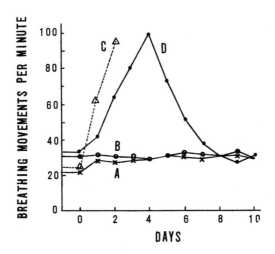

Figure 5.2 Average breathing frequency of four bluegill fish. Fish C and D were in tanks with a copper concentration of about 0.5 mg/l. A and B were in tanks containing unpolluted water. Fish D apparently recovered. The fish were in separate tanks.[1]

In contrast, a strong concentration of sulfuric acid, a potential water pollutant, has a conductivity of about 10^6 μmho/cm. One of the factors a physicist looks for in an environmental problem is a wide range of the variable being measured. This condition holds for conductivity measurements. If the water pollutant is known, as well as its conductivity, its concentration can be found from the changes in the conductivity of water.

There are hidden assumptions on which this simplicity is based. The linearity of the conductivity with concentration has been assumed. If x% of the pollutant produces an increase of y units of conductivity, then 2x% should produce an increase of 2y units.

This is true for the concentrations of most polluting chemicals. However, it is not true of all, and a counterexample is shown in Figure 5.3. It shows how the conductivity of natural water changes as P_2O_5 is added. The chemical reaction taking place is:

$$P_2O_5 + H_2O \rightarrow 2H_3PO_4 \tag{1}$$

The concentration of P_2O_5 in water in parts per million (ppm) is shown on the vertical axis of Figure 5.3.

These acids come from liquid fertilizer plants. The more that spills onto or filters into the ground, the worse the condition of nearby streams.

90 PHYSICS OF THE ENVIRONMENT

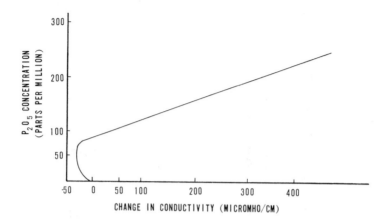

Figure 5.3 Change of conductivity of ordinary water due to the addition of phosphoric acid, H_3PO_4. Because of the chemical reaction represented by Equation 1, the concentration is measured in units of P_2O_5.[2]

In one fertilizer plant, instruments measuring conductivity were placed in drains and other receptacles to detect how much phosphoric acid (H_3PO_4) was leaking out. Estimated flowrates of contaminated water at each site are listed in Table 5.1.

Table 5.1 Determining the Minimum Conductivity Changes Around a Fertilizer Plant[2]

Location number	Estimated flow of contaminated water (l/min)	Concentration of P_2O_5 produced by 10 kg/day (ppm)	Change in conductivity to be detected, from Figure 5.3 (μmho/cm)
1	100	69	40
2	20		
3	5		
4	10		
5	550		
6	150		

Table 5.1 can be completed by finding the minimum conductivity change which can be measured by each instrument. Each site has at least 10 kg of P_2O_5 passing through per day. Now 1 liter/min of contaminated water corresponds to 1440 kg/day, since 1 liter equals 1 kg. The density of water is assumed to be unaffected by the addition of a slight amount

of contaminant. Then 10 kg/day corresponds to 0.0069 liter/min. There is then at least 0.0069 liter/min of P_2O_5 at each instrument. For example, 100 liter/min of total water flow at a given point produces a concentration of P_2O_5 of 69 ppm. In this way, the concentration of a water pollutant can be measured. This value can be compared to legal concentrations and leakage sources traced.

Many pollutants change the electrical conductivity of the water into which they flow. The method can then be used to track down a wide range of water pollutants.

Conductivity and Air Pollution

When pollutants are added to air, the change of conductivity is in the opposite direction to that in water. As the concentration of air pollutants increases, the conductivity falls. Since air has relatively poor conductivity, adding more particles to it with low conductivity decreases its overall conductivity.

Air conductivity does not usually change with pollutant concentration as in Figure 5.3. The relationship is shown in Figure 5.4. Note that x- and y-axes have been reversed in comparison to Figure 5.3.

The curve shows how the conductivity changes with the number of nuclei (or dust particles). The measurements of Figure 5.4 were taken on a mountaintop in Hawaii.

Measuring the conductivity of air is crucial in determining the extent of worldwide air pollution. Most pollution experiments are highly localized, measuring environmental conditions over a fairly small area. If we are ever to know how the world is faring under its increasing load of pollution, measurements in various parts of the globe are needed. Determining the conductivity of the air is one such measurement. The apparatus is simple. It can be used aboard ocean-going ships and is easily corrected for different atmospheric conditions.

Using this method to measure worldwide air pollution has another significant advantage. Starting in 1907, a series of global voyages was undertaken, in part to measure air conductivity. The data accumulated since then shows how air pollution has varied in different parts of the world for over a half century. Some of the information is shown in Figure 5.5.

In the two so-called "control areas," the South Pacific and North Atlantic, extensive measurements have been made over the years. Results are shown in Figure 5.6. In the North Atlantic, the conductivity is apparently decreasing with time, and in the South Pacific it seems to be fairly constant. This implies that the air pollution produced by North Americans

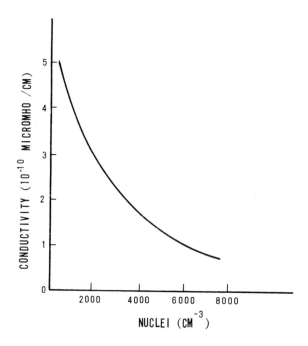

Figure 5.4 Relationship of atmospheric electrical conductivity to the concentration of nuclei—or dust particles—in the air. The measurements were taken on a mountaintop in Hawaii. They are an approximation to the conditions at sea level, at which height the worldwide data shown in Figure 5.5 were taken. The conductivity decreases with an increase in the number of dust particles.[3]

and Europeans is having significant effects in their hemisphere, but the southern hemisphere has escaped.

A simple quantity like conductivity can give us a good approximation of the extent of air pollution. Because this is one of the few environmental measurements which have been made for many years, it can be used to measure air pollution worldwide.

MEASUREMENT

Electrical measurements have been used in many branches of physics. They have only recently been applied to environmental physics, so the field is narrowed down somewhat.

ELECTRECOLOGY: ELECTRICITY AND MAGNETISM 93

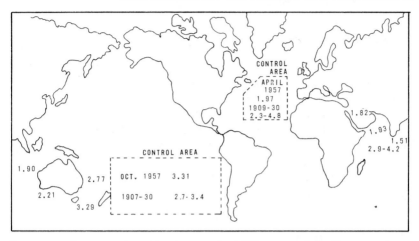

Figure 5.5 Some experimental points on the 1967 voyage of the *Oceanographer,* and its data compared to that of previous expeditions. The numbers indicate the atmospheric electrical conductivity in 10^{-10} μmho/cm. The "control areas" outlined in the North Atlantic and South Pacific are regions where the change in conductivity has been carefully investigated through the years.[3]

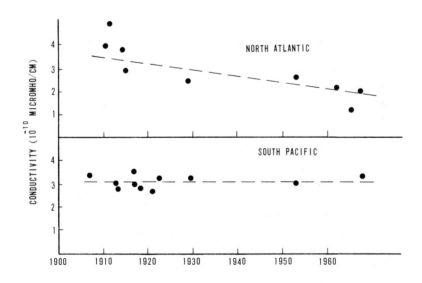

Figure 5.6 Change in atmospheric conductivity for the two control areas outlined in Figure 5.5. The conductivity in the South Pacific area has remained fairly stable, but that of the North Atlantic area has decreased drastically.[3]

Piezoelectricity and Air Pollution

Piezoelectricity occurs when certain materials have pressure applied to them and electric charges appear on their surfaces. This effect can be used to measure the levels of air pollution.

Piezoelectricity is found in the quartz that makes up our beaches. The effect is not quite as rare as might first be assumed. This piezoelectric effect also works the other way: if these materials are placed in an electric field, they twist or distort mechanically.

Suppose that instead of a constant electric field in piezoelectric quartz, there was an alternating field, of the form:

$$E = E_0 \sin \omega t \tag{2}$$

where E_0 = constant
 ω = the frequency of alternation of the field
 t = the time

When this changing electric field is applied to the quartz, the mechanical pressure it exerts on its holder will also change. The quartz will vibrate. The frequency of spring vibration is:

$$f = \frac{1}{2\pi} \sqrt{\frac{k}{m}} \tag{3}$$

where k is the spring constant, and m is the mass suspended at the end of the spring. How does this formula relate to our piece of quartz? The frequency of the applied electric field is varied until its resonant frequency is found. At this frequency, the quartz has one predominant motion. It behaves for all practical purposes like a mass on the end of a spring, governed by Equation 3.

Piezoelectricity may be put into context by recalling that one problem in measuring air pollutants is finding the mass, or concentration, of the particles in a given volume. A simple way, in widespread use today, is to draw enough air through a filter to deposit the pollution on it, and weigh the filter before and after. The process involves subtracting one large number from another—the weights of the filter paper before and after air is passed through it—to get a small number: the mass of the particles.

Equation 3 shows that the quartz vibration frequency is inversely proportional to the square root of the mass of the crystal. If the mass were changed by having airborne particles fall on it, then the frequency would change. The mass would change only slightly, but small frequency changes may be measured precisely. If the change in quartz vibration frequency is found, the mass of the airborne particles deposited on it can be determined. In practice, accuracy is much better than the filter paper method.

A schematic diagram of the system is shown in Figure 5.7. The particles pass through the metal electrodes and land on the quartz crystal. The change in frequency is measured.

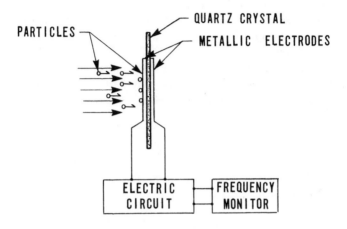

Figure 5.7 The piezoelectric system for measuring the mass of airborne particles. The metallic electrodes on the quartz crystal ensure that the air particles are directed to the right place. A frequency monitor detects small changes in the quartz vibration frequency.[4]

Suppose a mass Δm of air pollutants fall onto the quartz, where Δ indicates a small quantity. Then Equation 3 becomes

$$f + \Delta f = \frac{1}{2\pi} \sqrt{\frac{k}{m + \Delta m}} \qquad (4)$$

where Δf denotes a small change in vibration frequency. The object is to find the frequency change in terms of mass change. If top and bottom of the right-hand side are divided by m,

$$f + \Delta f = \frac{1}{2\pi} \sqrt{\frac{k/m}{1 + (\Delta m/m)}} \qquad (5)$$

Dividing out the factor k/m, the right-hand side is

$$\frac{1}{2\pi} \sqrt{\frac{k}{m}} \frac{1}{1 + (\Delta m/m)}$$

$$= \frac{1}{2\pi} \sqrt{\frac{k}{m}} \sqrt{1 - \frac{\Delta m}{m} + \left(\frac{\Delta m}{m}\right)^2 + \ldots}$$

96 PHYSICS OF THE ENVIRONMENT

The last expression was obtained by noting that a quantity $1/(1 + x)$, where x is very small, equals $1 - x + x^2 - x^3 + \ldots$, where the dots represent extra terms.

If $\Delta m/m$ is much less than one, then $(\Delta m/m)^2$ is considerably smaller and can be dropped from Equation 5 for not changing the results significantly. For example, if $\Delta m/m = 0.01$, then $(\Delta m/m)^2 = 0.0001$. $\Delta m/m$ itself cannot be dropped because there would then be no terms in Δm in the final answer.

Dropping some terms and keeping others constitutes comparing orders of magnitude, an important tool for the physicist. By using this reasoning, rough estimates can be obtained even when the exact numbers are not known. Then

$$f + \Delta f = f \sqrt{1 - \frac{\Delta m}{m}} \qquad (6)$$

where Equation 3 is used to find f. Using the binomial theorem, the right-hand side can again be written in powers of $\Delta m/m$. Then

$$f + \Delta f = f \left[1 - \frac{\Delta m}{2m} + \ldots \right] \qquad (7)$$

where the dots indicate terms in higher orders of Δm. Then

$$f + \Delta f = f - \frac{f \Delta m}{2m} \qquad (8)$$

Subtracting f from both sides and rearranging,

$$\Delta m = -\frac{2m \Delta f}{f}$$

The change in mass of the quartz as airborne particles fall on it is therefore proportional to f/m. Since f, the crystal resonant frequency, and m, its original mass, are both constant,

$$\Delta m = c \Delta f \qquad (9)$$

where c is a constant. Life becomes simple when one quantity changes linearly with another, rather than nonlinearly. The quartz crystal is calibrated by measuring a known change in mass as frequency varies a known amount, to obtain the constant c. A typical experiment is shown in Figure 5.8. The points are the values obtained when the crystal was weighed after air pollutants, consisting of tobacco smoke and ordinary particulate matter, were deposited on a crystal.

An example of typical readings in an industrial building is shown in Figure 5.9. The frequency of the crystal is measured and Equation 9 is used to go from frequency change to mass change. The resonant or

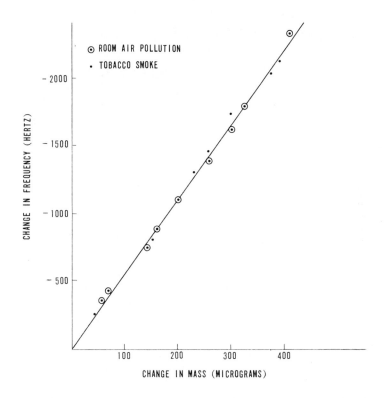

Figure 5.8 Mass change of a quartz crystal with the change in vibration frequency. The relationship is linear for both tobacco smoke and ordinary room air pollution. The latter comprises the particles found in most rooms.[4]

starting frequency is 1,599,500 Hz. The frequency drops much faster during working hours in the building, starting at time around 17 hr, than during nonworking hours, starting at times around 1 hr and 25 hr.

The results show that the air pollution during working hours is much more severe than that during nonworking hours. The maximum rate of mass deposition is reached just before quitting time. While the pollution levels after work is over are small, they are not zero. This indicates that the dust and smoke stirred up during the day take some time to settle.

Because the piezoelectric method allows many readings in a short time, the changing level of air pollution may be found quickly. This method has many industrial applications.

98 PHYSICS OF THE ENVIRONMENT

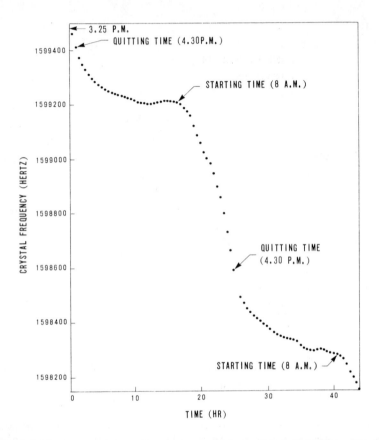

Figure 5.9 Frequency change of a quartz crystal in an industrial building. The frequency changes must be converted to mass changes to determine the levels of air pollution. The level is much higher during working hours than afterwards.[4]

Acid Rain

"Acid rain" (H_2SO_4), which eats at clothing, people and animals, has already fallen many times on Scandinavia. When coal and oil are burned, sulfur dioxide (SO_2) is produced. The SO_2 is oxidized (*i.e.*, another oxygen atom is added) in sunlight, in the reactions

$$\text{coal (carbon and sulfur)} + \text{oxygen} \rightarrow SO_2 + CO_2 + \text{energy}$$

$$2SO_2 + O_2 \rightarrow 2SO_3$$

The SO_3 in turn can combine with water vapor by the reaction

$$SO_3 + H_2O \rightarrow H_2SO_4$$

producing damage to lungs or even holes in clothing.

Sulfuric acid burns microscopic holes in plastic film as well as clothes. If electric current is passed through the film, the holes will act as tiny capacitors, storing charge, until they spark. A typical experimental set-up is shown in Figure 5.10. The values of the resistors and capacitor in the electrical circuit have not been shown, since they will depend on the thickness of the film and other considerations.

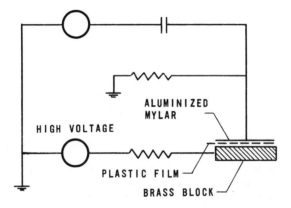

Figure 5.10 Circuit diagram of apparatus to measure concentration of sulfuric acid in air. Sulfuric acid falls on the plastic film, which is then inserted between the Mylar and the block. The current "jumps" across the gap, vaporizing part of the aluminum on the Mylar. The brass block completes the circuit. Resistors are indicated by a wavy line; capacitors by parallel lines.[5]

The experiment begins by the electrical sparks through the film striking another plastic film—the aluminized Mylar. The sparks cause the aluminum film on the Mylar to vaporize, leaving white spots which are big enough to be counted by the naked eye. The electric current allows magnification of the original holes.

Results of an experiment are shown in Figure 5.11. The spark count rises strongly with the sulfuric acid concentration. As a check on this method, the actual concentration of the acid was found using a Wilson cloud chamber, described in texts on atomic physics.

The electrical spark method can be used to measure the concentration of a particularly dangerous air pollutant, sulfuric acid, which is difficult to measure in other ways. Although there are limitations to the method, its use is fairly simple.

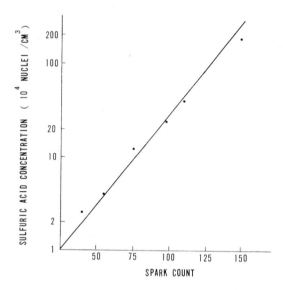

Figure 5.11 Spark count changes with the concentration of sulfuric acid particles. The spark count corresponds to the number of vaporized spots on the Mylar film. The concentration was measured independently on a Wilson cloud chamber. The relationship between the two variables is linear on semi-log graph paper, but may not be so for lower concentrations of sulfuric acid.[5]

The Size of Air Pollutants

There are at least two varieties of air pollution. One type is particulate matter, *i.e.*, solid matter. But some air pollution, like sulfuric acid rain, can be liquid. The rate at which they evaporate allows approximation of the particle size. In turn, knowledge of their dimensions gives a better understanding of their effect on human health. For example, large particles could be trapped in breathing passages, reducing their danger, while small ones may do the most damage.

Suppose electric current is passed through a thin wire. The wire becomes hot because of its resistance, and eventually reaches a stable temperature. If a drop of the liquid air pollutant falls on the hot wire, the metal is cooled, because the liquid is colder than the wire. Since the latter's resistance varies with its temperature, the resistance falls. At the same time, the drop of liquid is heated to evaporation. As it disappears, the wire resistance increases to its previous value. The process is repeated each time a drop hits the wire.

The electrical circuit is arranged so there is a constant current through it at all times. The voltage across the wire is given by Ohm's law, $E = IR$, where E is the voltage, I is the current and R is the resistance. Since I is a constant, the voltage will vary linearly with the resistance R.

Figure 5.12 shows a typical experimental result—an oscilloscope trace. The vertical scale is the voltage of the wire, and the horizontal scale is the time. The height of the spike depends on the voltage. To make it easier to read, the direction of the spike has been reversed.

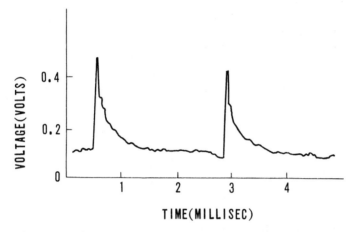

Figure 5.12 An oscilloscope trace recorded when a droplet hits an electrically charged wire. The voltage increases rapidly and then drops off gradually. The process is repeated when another drop hits the wire. The event ends within a few thousandths of a second.[6]

The voltage changes as predicted. The drop hits the wire just before each of the large spikes. As the drop is boiled away, the voltage rises quickly. The drop then disappears and the voltage returns to its normal level.

How is the size of the drop found? The simplest way is to use an energy balance. This calculation adds up all the energy going into a process and equates it with the energy going out. The former energy is the heat energy required to evaporate the drop of liquid. The latter energy is the electric energy required to produce the heat.

The heat energy is proportional to the mass m of the drop. It will also be proportional to the change in temperature the drop experiences. Other factors will be disregarded.

It is assumed that the drops are roughly spherical in shape. Then using the formula for a sphere's volume,

$$m = DV = \frac{4}{3}\pi r^3 D \tag{10}$$

where D is the density of the drop and V is its volume. The heat energy E_h is then, since D is a constant,

$$E_h \alpha m \alpha r^3$$
$$\text{or} \quad E_h = k_1 r^3 \tag{11}$$

where the constants have been combined into one constant, k_1.

It is assumed that the curves of Figure 5.12 represent exponential decay from a peak voltage change V_{max}, it can be shown that the electrical energy required to evaporate the drop is

$$E_e \alpha V_{max}$$
$$\text{or} \quad E_e = k_2 V_{max} \tag{12}$$

where k_2 is a constant.

Since the two energies are equal, the energy balance is then

$$k_1 r^3 = k_2 V_{max}$$
$$r = \left(\frac{k_1 V_{max}}{k_2}\right)^{1/3} \tag{13}$$

The drop radius then varies as the cube root of the maximum change in voltage of the wire.

Figure 5.13 shows how the size of the drop varies with voltage for two types of oil. Oils were used because they can easily be made into sprays. If Equation 13 is valid, a graph of $V_{max}^{1/3}$ versus r should be a straight line. This is approximately true for the two oils of Figure 5.13.

The voltage change of a wire when a liquid drop hits it can be used to measure the drop's radius. This method can be extended to find the size of real air pollutants, such as sulfuric acid.

Oil Slick Thickness

For spills on the high seas, knowing oil thickness is essential for determining the way a slick is spreading. Can electrical measurements be used to find this thickness?

One quantity, differing radically between oil and water, noted in Chapter 1, was the dielectric constant ϵ. Its value is about 80 for water and about 2 for oil.

This quantity can be used to measure oil thickness. For two parallel plates, the capacitance is

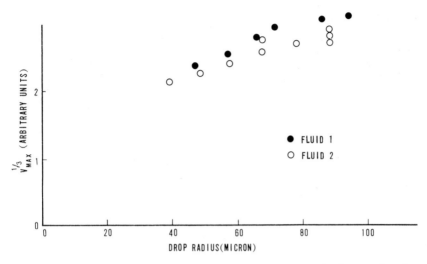

Figure 5.13 Radius of droplets falling on an electrically charged wire as a function of the maximum wire voltage change.[6]

$$C = \frac{\epsilon A}{d} \qquad (14)$$

where A is the area of the two plates and d is the distance between them.

The larger the dielectric constant, the more charge the capacitor can hold. This varies as its capacitance C. The larger the plates, the more charge they can hold, so C will vary directly with A also.

On the other hand, the farther apart the two plates are, the less influence the charge on one has on the other. This results in a smaller capacitance.

A schematic diagram of a typical experiment using this knowledge is shown in Figure 5.14. Parallel plates, one resting on top of the oil and the other in the water, are seen. We have, in effect, two capacitors in the electrical circuit—a "water capacitor" and an "oil capacitor." Because the current passes directly from one into the other, they are so-called "series capacitors."

The combination of two resistances in series is their sum:

$$R = R_0 + R_w \qquad (15)$$

where R_0 and R_w are resistance of the oil and water, respectively. Capacitances behave differently. To combine them, their inverses must be added:

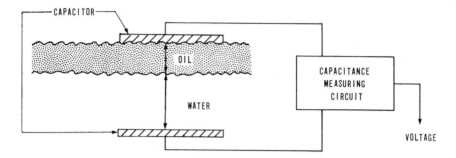

Figure 5.14 A parallel plate capacitor designed to measure the thickness of oil on water. The oil spill extends to the right and left.[7]

$$\frac{1}{C} = \frac{1}{C_0} + \frac{1}{C_w} \qquad (16)$$

where C, C_0 and C_w are the combined capacitance, the oil capacitance, and the water capacitance, respectively.

To derive Equation 14, parallel plates with a substance between them, not surrounding them, were assumed. Does the fact that the bottom plate is in the ocean change the physics of the situation?

While current flows between the two plates of Figure 5.14, and water conducts electricity, nearly all the current goes directly from one plate to the other. Almost none "leaks" into the ocean.

Experimentally, the voltage is measured. The voltage V between the two plates is defined as

$$V = \frac{K}{C} = K\left(\frac{1}{C}\right) \qquad (17)$$

where K is a constant. The term in brackets is Equation 16, so

$$V = K\left(\frac{1}{C_0} + \frac{1}{C_w}\right) \qquad (18)$$

Substituting values for C_0 and C_w from Equation 14,

$$V = K\left(\frac{t_0}{\epsilon_0 A_0} + \frac{t_w}{\epsilon_w A_w}\right) \qquad (19)$$

where t_0 = water thickness
t_w = water thickness
A_0 = oil plate area
A_w = water plate area

Since these two areas are equal, both equal A_0. Taking out the factor t_0,

$$V = Kt_0 \left[\frac{1}{\epsilon_0 A_0} + \frac{t_w}{\epsilon_w t_0 A_0} \right]$$

The same can be done for the factor $1/\epsilon_w A_0$, producing

$$V = \frac{Kt_0}{\epsilon_w A_0} \left[\frac{\epsilon_w}{\epsilon_0} + \frac{t_w}{t_0} \right] \quad (20)$$

The quantity to be measured, t_0, is in an equation containing a quantity we can measure, V. Assume that t_0 is about the same as t_w. Typical values for ϵ_w are about 80, and for ϵ_0 about 2. The first term in the brackets of Equation 20 will then be about 40 times the second term. Because the second term is much smaller than the first, it will be discarded. Then

$$V \cong \frac{Kt_0}{\epsilon_0 A_0}$$

The altered equality sign indicates the two sides of the equation are approximately equal. t_0 can then be found in terms of the other known quantities.

The capacitor plates are positioned so that t_w is about the same as t_0. Figure 5.15 shows that the experimental method is most applicable to relatively thick oil layers. There is a fairly linear relationship between the voltage V and the oil thickness t_0. The actual experiment was done in containers, with t_0 being measured independently. It is clear that this method of measuring thickness has the potential of good accuracy and reproducibility.

The concept of capacitance, or the storage of charge, may seem far removed from that of measuring oil thickness. However, the two concepts can be combined to find the size of an important menace to the environment.

Mining and Acid Wastes

Strip mines do more than rearrange earth. Acids from the disturbed earth often seep into nearby streams, ruining the rivers for recreational use. In Pennsylvania alone, it has been estimated that more than 18 billion liters of acidic water are produced each day—about 4% of the total flow over Niagara Falls daily. Much of this finds its way into adjacent streams. Exactly where are the sources of this acidic water? Many mines, both surface and undergound, have been abandoned through the years, and finding them is often difficult.

106 PHYSICS OF THE ENVIRONMENT

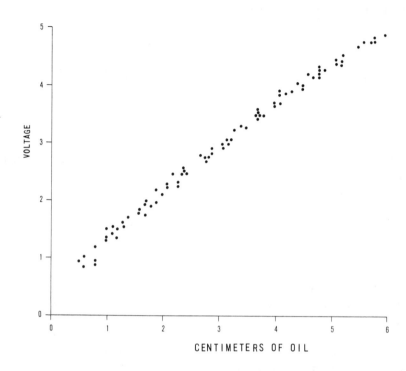

Figure 5.15 Thickness measurement of oil on water using the capacitance circuit of Figure 5.14. Although there is some scatter in the readings, the line is fairly straight. The thickness of the oil was measured independently.[7]

It has been shown that measuring the resistivity of polluted water gives a measure of the degree of pollution, since the resistivity decreases with increasing pollution. The same principle holds in mine water. Figure 5.16 shows how the resistivity varies with the ion concentration, or degree of acidity.

Consider a real case, observed near Kylertown, Pennsylvania. In explanation, the resistivity in Figure 5.17 was measured not only at the surface, but below it too, since water flows through the ground as well as on top. The northern surface area has considerably lower values of resistivity than any other region or depth, indicating that the ion concentration is highest there. Mine acid is probably draining in that area.

Although there are significant problems in applying the water resistivity method to find where mines drain underground into streams, the results of Figure 5.17 give hope.

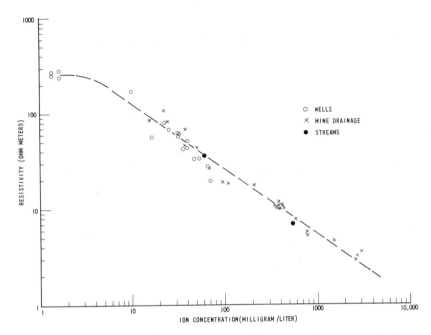

Figure 5.16 Water resistivity changes with ion concentration. The concentration is a measure of the quantity of pollution in a unit volume of water. Wellwater has a higher resistivity, generally speaking, than mine drainage water.[8]

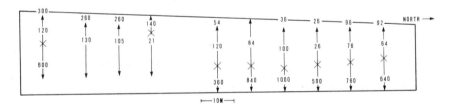

Figure 5.17 Resistivities (in ohm-meters) of mine acid near Kylertown, Pennsylvania. The figures for resistivity below the surface are averages over distances of the order of 15 m. The resistivity—and thus the ion concentration—varies considerably from place to place under the soil, giving a clue as to where and how fast the acids from the mine are draining.[8]

Microwaves on the Ocean

The information which goes into making the picture on a television screen is transmitted by means of electromagnetic radiation. These signals are the basis of most long-distance communication in the world today.

108 PHYSICS OF THE ENVIRONMENT

Radiation can be used to detect oil spills. If electromagnetic radiation is beamed at the ocean, a certain type of signal returns. If it is beamed at an oil-covered portion, another type of signal is received. The oil reflects the radiation in a different way.

A typical experimental situation is shown in Figure 5.18. An aircraft flies over a region where a spill is suspected, and instruments on board emit electromagnetic radiation. On another aircraft, other instruments pick up the reflected signal. If necessary, both the emitter and receiver could be carried on the same aircraft. This method allows a tremendous range of area.

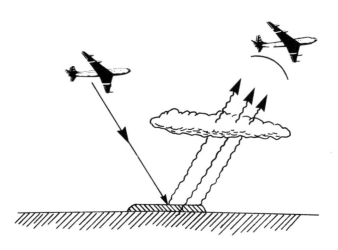

Figure 5.18 Finding the location of oil slicks by emitting electromagnetic radiation from an aircraft. The changes in the characteristics of the radiation after it bounces off the surface of the sea is recorded by a second aircraft, and correlated with known changes for previously measured oil slicks. If the beam is sent directly downward, the same procedure could in principle be carried out with one aircraft.[9]

The difference in reflection ability (or reflectivity) that the receiving instrument measures can be described as a "change of brightness temperature." Over the ocean, the reflected radiation is different from the initial radiation in such factors as its signal strength and its degree of distortion. These factors are combined mathematically into the brightness temperature. The actual temperatures of the oil and the oceans are the same. But the difference in their brightness temperature—or change of the electromagnetic signal—is not zero.

Figure 5.19 shows the brightness temperature for some oil spills. The waves reflect the radiation in a nonuniform way, and this is shown in the results. Where an oil slick is encountered, the brightness temperature rises drastically. As an example, for wavelengths of 3.2 mm the brightness temperature goes up from about 30°K to 90°K.

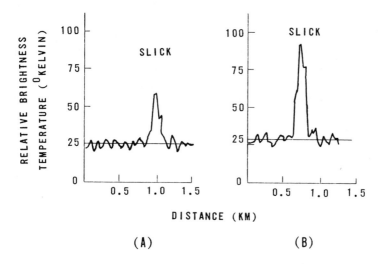

Figure 5.19 Change in the "brightness temperature" as an aircraft flies over an oil slick. This temperature is a measure of the change in the characteristics of the electromagnetic radiation reflected from the surface of the sea. The graph shows clearly where the slick lies. Diagrams (A) and (B) refer to wavelengths of electromagnetic radiation of 8.1 and 3.2 mm, respectively.[9]

Finding oil slicks by electromagnetic radiation is only one of the many measurement and control methods which deal with that substance.

Ocean Currents

The way civilization's waste spreads from one pole to the other is due, in part at least, to the sea currents that carry it along. Probably the most famous of these currents is the Gulf Stream, flowing from America to Europe. One of the greatest problems in studying these ocean movements is the lack of adequate instruments to tell just how fast and in which direction the sea moves.

Electricity can be used to measure slow sea currents. An electrical resistor gives off heat and is cooled by flowing water, the same technique as finding which way the wind blows by holding up a wet finger.

By Joule's law, the thermal power generated in a resistor is

$$P = I^2 R \qquad (21)$$

where I is the current and R the resistance. If the resistor is placed in the ocean, it cools off. It cools even more as the water flows by, since the water carries away the heat faster. This extra cooling is correlated with the rate of flow.

In practice, the resistor temperature is kept constant by a thermometer and electronic devices, and the extra electrical current I needed to keep the temperature constant as the ocean flow varies is measured. Figure 5.20 shows how the extra electrical current passed through the resistor to maintain its temperature changes with ocean flow. The electrical current increases rapidly with flow at low velocities, but tends to level off at higher velocities.

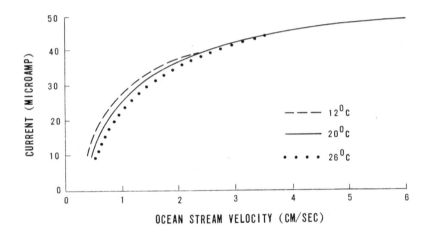

Figure 5.20 Current required to bring a standard electrical resistor back to its original temperature after it is placed in moving water. Slightly different results are obtained depending on the temperature of the water. The effective range of the instrument in terms of ocean stream velocity is up to about 5 cm/sec.[10]

A slightly different curve is obtained depending on the temperature of the water. When its temperature is increased somewhat, it takes less heat to bring the resistor back up to its "standard" temperature, so less electrical current is required.

This method measures the velocity of ocean and lake currents over a wide range. When these velocities are known, sea pollution can be studied in a more rational manner. The way wastes flow in the sea can then be found, as well as the pollution patterns of coastal cities and oceanside industrial polluters.

Polarography

Polarography is a method for finding the level of very small concentrations of pollutants in water. The process is a cross between physics and chemistry. Suppose part of the wire in an ordinary circuit is replaced by a glass tube containing mercury. Since liquid mercury conducts electricity, there is still a working circuit. The liquid mercury can be mixed with pollutant-containing liquids, which change the electrical characteristics of the circuit. The property measured by this process is the electrochemical potential. Every pollutant ion has a different electrochemical potential. If the same voltage is applied to different ions, different currents move through the circuit.

The electrochemical potential of different elements can be used to find the concentration of such water pollutants as copper, cadmium and zinc in the part per million range. In Figure 5.21, the polluted water is in the container labelled A. The mercury is enclosed in the long tube labelled B. The water is fed into the mercury, and the mixture drops slowly, drop by drop, into the pool of mercury in the container C below. The circuit is completed by the right-hand electrode D.

At a low voltage, the current increases very slowly, as shown in Figure 5.22. As the voltage increases from point B to C, the current rises sharply. There are now enough ions of the pollutant (cadmium in this case) in the electrical system to change the current drastically. When the voltage becomes relatively high, in the region from point C to D, pollutant ions move more slowly through the system, and so the current levels off.

The midpoint of region BC in Figure 5.22 has a voltage of -0.68 volts (V), which is related to the electrochemical potential voltage. Each pollutant has a characteristic midpoint voltage which can be used to identify it. In this example, -0.68 V would be characteristic of cadmium.

One way of determining the midpoint more accurately is not to plot current vs voltage, but rather the slope of this curve. In Figure 5.22 the slope would be small over regions AB and CD, and large over region BC. The midpoint of the high-slope region would be the characteristic voltage.

In Figure 5.23, the liquid contains different metallic pollutants. There is a characteristic voltage at which each kink occurs. By using this method, the occurrence of copper, cadmium, nickel and zinc in water can be found in sequence. Part of the graph has been expanded, showing tiny oscillations as the mercury droplets fall. These oscillations are too small to be seen in the main diagram on the left.

Polarography is a potentially powerful method for detecting minute traces of pollutants in water. Because of its simplicity, it is being used more and more.

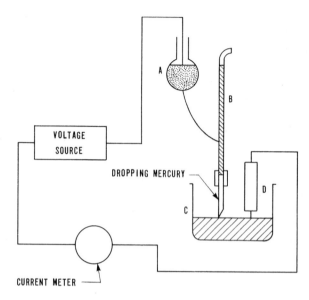

Figure 5.21 Polarograph for measuring the presence of trace metals in water. The polluted water is in container A. It passes into tube B, which contains mercury. The mercury drops into container C, and the current in the circuit passes through the electrode D. The change in current as the voltage changes in the circuit shows the presence of certain pollutants in the water.[11]

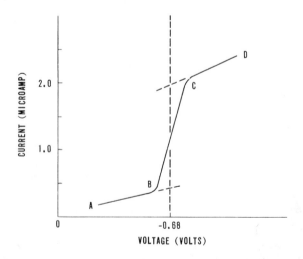

Figure 5.22 Typical polarogram—or graph of current vs voltage—for cadmium in water.[12]

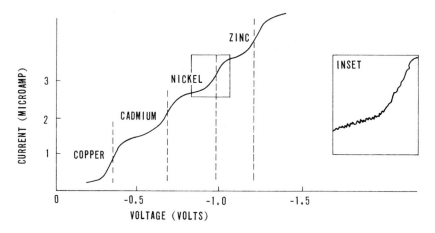

Figure 5.23 Simultaneous determination of copper, cadmium, nickel and zinc in water by polarography. Each of the elements has a characteristic voltage at which the current rises rapidly. The inset shows the detail of the rise for nickel.[12]

CONTROL

Filtering Water

Filtering out impurities is one way to keep water clean. If a fluid has any solids in it, the solids gradually build up in a layer on the filter. This layer blocks the fluid from passing through the filter, and eventually nothing passes through.

The problem of buildup of a solid layer is inherent in all water filters. One way to prevent this is by using electricity. Most naturally occurring substances in water carry a negative charge. For example, clays are negative electrically because certain atoms lacking a positive charge substitute for others in the regular clay structure. Similarly, bacteria and algae have negative charges on their cell walls. These facts may be used to get rid of the accumulated solids which soon block most filters.

In Figure 5.24, an electric current is passed through the filter, shown in the center of each of the three diagrams. In diagram B, there is no voltage on the filter, so the cake blocks the flow of liquid. The two membranes do not play any part in filtering the liquid, but only hold the solids in captivity.

In diagram C, a small negative voltage has been applied to the filter. Since the solids are negatively charged, as mentioned above, they are repelled from the filter.

114 PHYSICS OF THE ENVIRONMENT

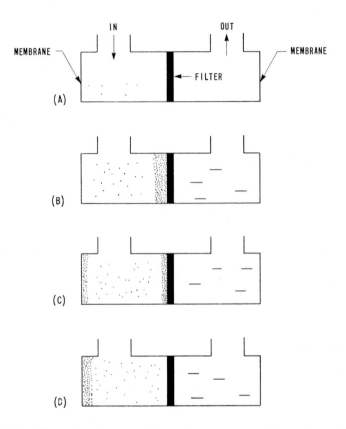

Figure 5.24 Electrophoresis, or applying voltage to a water filter to loosen the accumulated pollutants. (A) Schematic diagram of flow system. (B) Initial conditions, with no voltage applied. Solids build up on the filter and prevent the solution from flowing through. Dots represent solids. (C) A small voltage is applied. Some of the solids are repelled to the membrane on the left. (D) Voltage is increased high enough the remove all solids from the filter.[13]

In diagram D, enough voltage has been applied to the filter to repel all the particles to the membrane wall, and the filtering action goes on unimpeded. By applying a voltage, a larger volume of fluid flows through than would have otherwise.

As time goes on, solids will fill the left-hand compartment. The voltage on the filter is removed for a few moments, so that the solids disperse through the liquid in the left-hand compartment. A valve—not shown—drains the liquid. The entire process is electrophoresis, or the movement of charged particles in a fluid subject to continuous electric field.

There are two aspects of the process which are easy to measure, and are important in understanding how effectively it works. First is the volume of water, V, passing through the filter, and second is the time, t, it takes.

The fraction t/V is a function of V, the volume, and E, the voltage applied to the filter. In Figure 5.25, the lines indicate five different values of voltage E, ranging from 0 (line A) to 12 V (line E). Since V should be as large as possible, E is varied until the maximum value of V is obtained. The higher the voltage, the lower the value of t/V. In turn, this means a higher value of V. A high voltage should be applied.

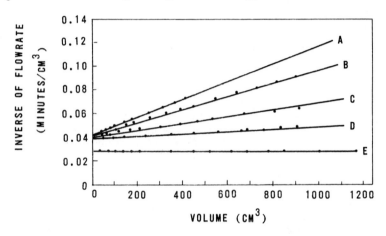

Figure 5.25 Inverse of the flowrate (t/V) through an electrified filter, as a function of the accumulated volume V of solution. Lines A, B, C, D and E correspond to 0, 2, 4, 6 and 12 V, respectively.[13]

As civilization produces more and more wastes, filtration systems must become more and more efficient. Applying voltages to filters offers a way to reach that goal.

Electrostatic Filters

Electrostatic precipitation is one of the most important devices for eliminating air pollution. It works on the principle mentioned in the last section—charged particles repel like and attract unlike. Figure 5.26 shows a simplified version of a typical precipitator. Dust-laden air enters on the bottom left-hand side. The object in the center is the precipitator itself. It consists of a hollow tube which is positively charged or grounded, giving the tube either a positive or a zero potential, plus a thin wire at the center of the tube which is negatively charged. A large number of positively and negatively charged ions are produced in the air, and cluster around the wire. The negatively charged particles stream away from the negative wire, and become attached to the neutral dust particles in the tube. The dust in turn becomes negatively charged. It then travels to regions of positive or zero potential. The dust strikes the outer tube and falls into the collector. The now-clean air moves out the top of the tube.

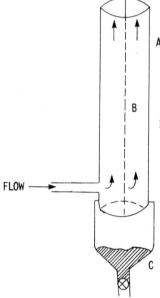

Figure 5.26 Arrangement of a cylindrical electrostatic precipitator. Tube A is the outer electrode, grounded or electrically positive. Wire B is the central electrode, strongly negative. C is the dust collector.[14]

Because of the great difference in potential between the wire and the tube, there is a strong electric field between the two parts of the precipitator. If the field is not powerful, the dust particles do not become charged—or ionized—and the dust moves out the top.

Figure 5.27 shows how the efficiency varies with applied voltage and the velocity of the airstream. The efficiency is defined as the percentage of dust particles which are removed from the air as it passes through the precipitator. It can easily be greater than 95%. Many precipitators have efficiencies of greater than 99%, making them the most effective anti-pollution devices known.

The two lines in Figure 5.27 show experimental data for airstream velocities of 1.0 m/sec and 1.5 m/sec. One would expect more dust to be taken out of the air at the lower velocity, since this means that the dust is in the precipitator longer, and so the electric field can work on

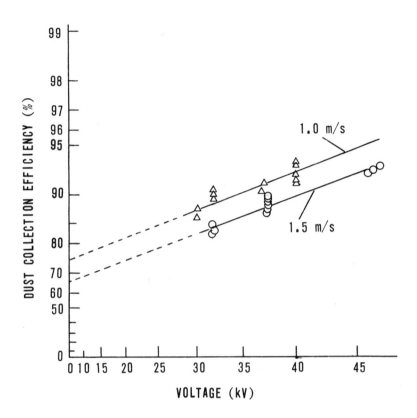

Figure 5.27 Relationship between efficiency, voltage and velocity of airstream through an electrostatic precipitator.[14]

it a greater length of time. However, this implies that it takes longer to clean the air.

There is more to a precipitator than just turning on an electric field and taking dust out of the air. Voltages, efficiencies, and the rate at which we pass the air through the machine must be considered. However, by a judicious combination of these factors large amounts of air pollution can be eliminated.

SUMMARY

Electricity and magnetism are the basis of a considerable proportion of physical measuring instruments. In turn, these instruments can be used to describe a wide variety of environmental conditions. The properties of electricity itself can be used to eliminate pollution in the air as well.

PROBLEMS

5.1 Measure the position of the spikes in Figure 5.1. Is the fish breathing regularly? Would it be better to use the small spike just after the large one for computations? What accounts for any irregularity in data? On the average, how many opercular movements does the fish make per minute?

5.2 What happened initially to the copper-polluted fish in Figure 5.2? Why did the rate of breathing for D eventually go down? Does the graph show "recovery" or not? Discuss a method for showing the pollution suffered by fish C and D. Hint: think of the cumulative dose.

5.3 Why is there such a large difference in conductivity between distilled and ordinary water?

5.4 Suppose ordinary water has a conductivity of 180 units, and ingredient X generally increases it by 20 units for each 1% concentration in water. What would a measurement of 3,000 units of conductivity show for a mixture of X and water?

5.5 Based on Figure 5.3, approximately what range of concentration of P_2O_5 in water would be difficult to measure? How would concentrations in this range be measured? Over what range does the concentration change linearly with conductivity?

5.6 Complete Table 5.1 by calculating the concentration of P_2O_5 produced by 10 kg of this pollutant per day. Then calculate the change in conductivity measured as a result of this concentration, using Figure 5.3. Extrapolations of the linear part of Figure 5.3 may have to be made. Are these extrapolations valid physically?

5.7 Discuss some of the limitations in detecting water pollutants by electrical conductivity. In particular, consider two pollutants in the water at the same time.

5.8 The resistivity of air is 1/conductivity. Does the resistivity increase linearly with the number of air particles? Calculate the resistivity from Figure 5.4 for air particle densities of 0, 1,000, 2,000 . . . 8,000 per cm, and plot it. Is there a linear relationship?

5.9 One way of narrowing the degree of uncertainty in Figure 5.6, for those acquainted with statistics, is to calculate the equations of the two lines, each giving point equal weight. Find the standard deviation of the slope of the two lines. This is one way of proving that the conductivity in the North Atlantic control area is decreasing.

5.10 Assume that the lines drawn in Figure 5.6 are the best fit to the experimental points. Then find the equations of the two lines. The conductivity can be written as a function of the quantity called "years since 1900."

5.11 In the North Atlantic area, when will the conductivity of the air be 10^{-10} μmho/cm?

5.12 Assuming that the relationship between conductivity and the number of particles in the air is given by Figure 5.4, plot the number of particles in the air per unit volume over the North Atlantic since 1900. Is it increasing linearly?

5.13 Suggest physical reasons why air pollution apparently is increasing in the North Atlantic and remaining constant in the South Pacific. Hint: where are the main sources of air pollution? Determine how weather conditions can affect where pollutants fall.

5.14 Determine numerically whether or not the approximations made to obtain Equation 9 are valid. Substitute $f = 1$, $k = 197$, and $m = 5$ into Equation 4, temporarily not using units. These quantities fit Equation 3. Let $\Delta m = 0.2$ and 0.5. What are the values of Δf obtained from the two equations?

5.15 From Figure 5.8, find the value of the constant c in Equation 9.

5.16 Using the data of Figure 5.9, how much mass was deposited on the crystal during the first 15½ hours of nonworking time? During the second? Is the difference significant? Around what time does the greatest rate of air particle deposition occur? What is the rate per hour at that time?

5.17 Describe physically what is happening to the crystal in Figure 5.9 around the time 12 hr and 37 hr. Is this a possible drawback to this method of measuring air pollution? Determine the average rate of air particle deposition during working and nonworking time.

5.18 In Figure 5.11, what would be the concentration of sulfuric acid at which no spark counts would be recorded? Is this a possible limitation on the method? What would be the concentration if 200 counts were recorded? How could the sensitivity of the method be improved? Hint: think about the size of the electrical sparks through the film. What are the general limitations of this method?

5.19 What is the maximum increase of voltage for the two spikes in Figure 5.12? How long does it take for the increase to drop to one half its maximum for each of the two cases? To one-fifth its maximum? Based on these numbers, are the two drops which struck the wire the same size? Plot the data of Figure 5.12 on semi-log graph paper, with the voltage on the vertical axis and time on the horizontal axis. Discuss the differences between the shape of these curves and those of Figure 5.12.

5.20 In the derivation leading to Equation 13, explain how the fact that drops are not always the same size could affect the results.

5.21 Based on the evidence of Figure 5.13, is Equation 13 valid? Over what range could the voltage of the wires to measure the drop radius be used? What is the value of the slope of the line in each case? Is the difference significant? Is the method useful for values of r under 30 microns (μ)? Is there a physical reason why?

5.22 Show that the combined capacitance is always less than either of the values of the oil and water capacitance. Do this either algebraically or by means of a numerical example. The combined resistance is always greater than either value of the oil and water resistance.

5.23 Using Equation 20, plot the value of V as the proportion of oil to water thickness varies from 1 to 0.1. Let the value of the constants in front of the brackets equal 1, for simplicity. This curve shows the range of validity of the assumption that t_o approximately equals t_w.

5.24 What difficulties in the capacitance method for measuring oil thickness are there in applying it outdoors?

5.25 Does the line drawn through the points of Figure 5.15 pass through the origin? Why or why not?

5.26 Based on Figure 5.15, can the capacitance method be used for very thin films of oil (less than 1 mm) on water? Why or why not?

5.27 What is the range of resistivity and ion concentration for groundwater in Figure 5.16? Answer the same question for mine drainage water.

5.28 For each of the resistivities in Figure 5.17, find the ion concentration from Figure 5.16. Disregard the individual slightly scattered points in Figure 5.16. Indicate roughly the resistivities along the different layers for both the north and south regions. About how much greater is the ion concentration for the top layer in the higher region than in the lower region?

5.29 What are potential problems in applying the resistivity method of locating mine drainage sources?

5.30 Detecting oil spills by electromagnetic radiation is similar to detecting aircraft by means of radar. Write a short description of the principles of radar. By your description could an old-time wood aircraft be distinguished from a metal one? While oil is not as different from water, from the point of view of radiation, as metal is from wood, the same principles of differentiation apply.

5.31 What is the approximate percentage increase in brightness temperature for each of the two peaks in Figure 5.19? Using the distance scale on the bottom of the graphs, what is the approximate diameter of the slicks? How does the range—the distance from minimum to maximum—of the slick areas compare to that of the nonslick areas? Could this be used as a criterion for the peak?

5.32 Based on Figure 5.19, which of the two wavelengths should be used for detecting oil slicks?

5.33 Is there a physical reason why the current measuring system in Figure 5.20 seems to fail below about 0.3 cm/sec?

5.34 At what velocity is the instrument of Figure 5.20 most sensitive to small changes in velocity? How much more sensitive is it at 1 cm/sec than 5 cm/sec?

5.35 Determine the average velocity of the Gulf Stream and find what current the instrument of Figure 5.20 would use to measure it.

5.36 Which temperature in Figure 5.20 has the highest sensitivity to changes of velocity at 0.3 cm/sec? Explain.

5.37 At what velocity is the difference in water temperature in Figure 5.20 affecting the current insignificantly? Judging when two or more lines come close enough together to be called "reasonably identical" is an important part of a physicist's craft.

5.38 Is the ocean flow-measuring instrument useful above about 4 cm/sec? Explain.

5.39 Which ocean velocity in Figure 5.20 has the largest difference in electrical currents for the three temperatures? What is the percentage difference at that point between the highest and lowest temperatures?

5.40 Does the current-voltage curve in Figure 5.22 pass through the origin? Why or why not?

5.41 How would the current in Figure 5.22 vary if the voltage were increased beyond point D? Would it make any difference in the results?

5.42 Suppose the slope of BC in Figure 5.22 were low. Would this make it more difficult to obtain an accurate result? Why?

5.43 Find and tabulate the midpoint voltages for the four elements shown in Figure 5.23. Look up the electrochemical potential for each.

5.44 As a result of negative charges in water, could the entire earth have an excess of negative charge? Could there be some parts of the world with an excess of positive charge?

5.45 Consider the volume V of Figure 5.25 at a few values, such as 200 cm^3 and 400 cm^3. Each has a corresponding value of t/V, depending on the voltage E. Plot V/t, the rate of flow vs E for each of the selected values of V. Which voltage produces the greatest fluid flow?

5.46 What is the ratio of the highest flowrate to the lowest for each value of V chosen for Problem 5.45? Does V/t change linearly with E? Answer the two questions for each value of V you have chosen. Could the answer to the second question be guessed from Figure 5.25?

5.47 What would be a practical problem in trying to make the inner electrode of a precipitator more and more negative?

5.48 Would an electrostatic precipitator produce the same results if the outer electrode were rectangular in shape?

5.49 What are the efficiencies at the two velocities of Figure 5.27 at a voltage of 55 kV? The scale of the x-axis increases with the square of the voltage.

5.50 Would the radius of the tube of an electrostatic precipitator affect the electric fields or the efficiency?

5.51 Plot a hypothetical curve of precipitator efficiency vs velocity. Think carefully about the regions near zero velocity.

5.52 It seems clear that the efficiency of a precipitator goes up as velocity decreases. What would be a reason for using high velocities?

REFERENCES

1. Spoor, W.A., T.W. Neiheisel and R.A. Drummond. "An Electrode Chamber for Recording Respiratory and Other Movements of Free-Swimming Animals," *Trans. Am. Fish. Soc.* 100(1):26,27 (1971).
2. Corrigan, P.A., V.E. Lyons, G.D. Barnes and F.G. Hall. "Conductivity Measurements Monitor Waste Streams," *Environ. Sci. Technol.* 4(2):119 (February 1970).

3. Cobb, W.E., and H.J. Wells. "Electrical Conductivity of Oceanic Air and Its Correlation to Global Atmospheric Pollution," *J. Atmos. Sci.* 27:815, 817, 818 (August 1970).
4. Olin, J.G., and G.J. Sem. "Piezoelectric Microbalance for Monitoring the Mass Concentration of Suspended Particles," *Atmos. Environ.* 5(8):654, 662, 666 (1971).
5. Stickney, J.E., and J.E. Quon. "Spark Replica Technique for Measurement of Sulfuric Acid Nuclei," *Environ. Sci. Technol.* 5(12): 1212, 1215 (December 1971).
6. Goldschmidt, V.W., and M.K. Householder. "The Hot Wire Anemometer as a Aerosol Droplet Size Sampler," *Atmos. Environ.* 3:646, 648 (1969).
7. Sauer, W.E., and V. Klemas. "Oil Layer Thickness Monitor," *Adv. Instr.* (Instrument Society of America) 26 (IV):840-3 (1971).
8. Merkel, R.H. "Use of Resistivity Techniques to Delineate Mine Acid Drainage in Ground Water," *Ground Water* 10(5):39, 41 (1971).
9. Edgerton, A.T., D. Meeks and D. Williams. "Microwave Emission Characteristics of Oil Slicks," presented at Joint Conference on Sensing of Environmental Pollutants, Palo Alto, California, November 8, 1971, paper 71-1071, pp. 1, 4.
10. Bartoli, G., G. Sandrelli and G. Sorrentino. "Instrument to Measure Weak Sea Currents," *Rev. Int. Oceanog. Mediteranean* XVII:114 (1970).
11. Mancy, K.H. *Instrumental Analysis for Water Pollution Control* (Ann Arbor: Ann Arbor Science Publishers, Inc., 1971). p. 92.
12. Johnson, B.R. "Use of Polarography in Water Pollution Control," *Water Poll. Control* 70(6):619, 621 (1971).
13. Moulik, S.P. "Physical Aspects of Electrofiltration," *Environ. Sci. Technol.* 5(9):771, 774 (September 1971).
14. Potter, E.C. "The Electrofiltration of Particulates from Gases," in *Electrochemistry of Cleaner Environments* (New York: Plenum Press, 1972), pp. 132, 138.

CHAPTER 6

LIGHT AND OPTICS

"Thus black and white and every other color will appear to us as produced by the encounter of our eyes with something which moves in the direction of the eyes."
                                    Empedocles (Greek), 5th century B.C.

". . . All rays of light have the same speed."
                                    Euclid, *Optics*.

". . . when (light) reaches a mirror (it) is not allowed to penetrate nor is it allowed to stop there and, since it still has the force and nature of its original motion, it is reflected on the side from which it came, along a line which has the same inclination as the first line."
                                    Alhazen (Egyptian), 10th century

"My design in this Book is not to explain the Properties of Light by Hypotheses, but to propose and prove them by Reason and Experiments. . ."
                                    Newton, *Opticks,* 1704

INTRODUCTION

The first three quotations might lead one to believe that the major problems of optics had been solved by the 10th century. The first deals with vision, the second with the constant velocity of light and the third lays the basis for geometrical optics, one of the most important branches of the subject. Around the time of Newton, experimentation gradually took over from speculation.

Light and Water

Because so many phenomena in the environment give off electromagnetic radiation in the form of light, optics is a powerful tool.

Table 6.1 illustrates the use of optics in just one aspect of the environment—water. The top row indicates possible sources of water pollution. On the left are listed the ways the problems are measured or observed.

Table 6.1 Using Optical Effects to Measure the Environment[1]

	Oil Spillage	Algae	Water Productivity	Tracking of Discharges	Municipal Sewer Discharge
Petroleum Spill	X[a]				
Visible and Ultraviolet	0[b]				
Chlorophyll Content		X	X		X
Reflection in Infrared		0	0		0
Bioluminescence			X		
Wave Height				X	
Sun Glitter				0	
Ocean Color				X	
Turbidity and Clarity				X	
Visible Variations				0	
Currents (speed and direction)				X	
Visible				0	
Surface Reflective				0	
Surface Contaminants (Foam, petroleum, fish oil, debris)				X	X

[a] X denotes major categories of measurement.
[b] 0 represents subcategories.

In Table 6.1 "X" denotes major measurement categories, and "0" denotes subcategories. For example, the chlorophyll content of open water is an important factor in measuring its biological productivity, as well as in tracking municipal sewage discharges. To find this content, reflection in the infrared portion of the spectrum can be used. Similarly, wave heights may be measured indirectly by charting the type and duration of sun glitter off the waves.

Tables like this one suffer somewhat from their generality. However, they do indicate the wide diversity of environmental measurements possible.

Asbestos Detection

Air pollution, oil slicks and thermal pollution all occur outside. But environmental degradation does occur inside our homes, offices, classrooms and factories. The most dangerous indoor environments are often found in workplaces, since it is there that noxious materials are hammered, mixed chemically and sprayed.

Asbestos, one of the best heat insulators known, is used in many products, ranging from brake linings to oven mitts. When asbestos is sprayed through the air, some can land on the workers doing the spraying. If any asbestos gets into the lungs, it can pose a serious health problem.

To identify asbestos in the air, factory dust could be collected and the components distinguished by chemical means. This, however, would be a slow difficult process because of the tiny concentrations involved. A simpler way would be to measure the optical properties of asbestos dust, and compare them to the optical properties of dust that do not contain asbestos. If the two sets of properties are different, then a method of detecting this dangerous substance exists.

Examples of optical measurements of transmittance on three types of asbestos are shown in Figure 6.1. If light of a particular wavelength passes through a substance, a certain amount will be transmitted; the rest will be absorbed or reflected. If the reflectance is very small, this following formula holds:

$$T + A = 1 \qquad (1)$$

where T is the light intensity transmitted, and A is the intensity absorbed. On the right-hand side, unit light intensity has been assumed. The light intensity transmitted or absorbed depends on the atomic and molecular structure of the substance.

Figure 6.1 shows that the transmittance (T) varies drastically for even slight changes in **wavelength**. Different wavelengths of light excite the electrons in the atoms in different ways, producing different types of absorption.

The wavelengths used in this experiment are in the infrared region. This range of wavelengths was chosen because there are characteristic changes in transmittance for this particular range.

For all three types of asbestos, there is a characteristic drop in transmittance at around $10\,\mu$. The transmittance gradually rises from around 40% at 3×10^4 Å. At the longer wavelengths, there is some difference between the types of asbestos.

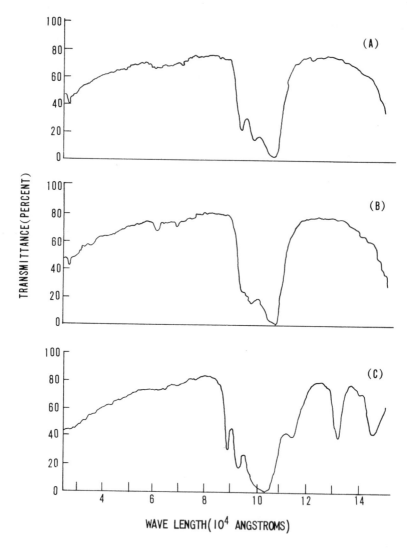

Figure 6.1 Variation of the transmittance of three types of asbestos with increasing wavelength of infrared light.[2]

The existence of asbestos and other air pollutants can be found by evaluating their transmission characteristics as a function of wavelengths. Interpretation of the transmission curves takes skill because of their detailed nature. However, the optical method is one of the simplest and most reliable available.

Underwater Light Absorption

Consider kelp plants, a type of seaweed, growing in water. If the water is clear, they can be seen clearly. If it is murky, they are seen dimly or not at all. This is a method of detecting cloudy water which can be useful in situations where more sophisticated techniques are impractical. For example, this method could be important if large water areas were to be monitored.

The physical situation is shown in Figure 6.2. In Equation 1, reflectance was disregarded as it played only a small part in the experiment on asbestos. In the outdoors, reflectance is often important. Equation 1 then becomes

$$T + A + R = 1 \qquad (2)$$

where R is the light intensity which is reflected. The left-hand side means that all light is transmitted, absorbed or reflected.

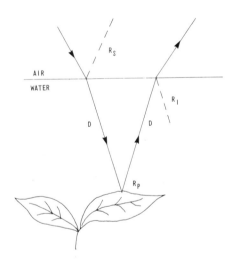

Figure 6.2 Decrease of the intensity of light as it reflects off underwater foliage. A fraction R_S of the original intensity is reflected off the air-water surface, a fraction R_p off the foliage and, finally, a fraction R_I from the underside of the air-water surface.[3]

In Figure 6.2, a unit intensity light falls on the surface of the water. If intensity R_s is reflected from the surface, then $(1 - R_s)$ is left to travel through the water. The Beer-Lambert law states that "the fractional change in intensity of light over a given distance is proportional to that distance." This can be written

$$\frac{\Delta I}{I} \alpha \Delta x \qquad (3)$$

where I = the intensity
       x = the distance
       Δ = small change.

Because light always decreases in intensity as it travels, *i.e.*, ΔI is negative, the proportionality constant must be negative. Then

$$\frac{\Delta I}{I} = -K \Delta x \qquad (4)$$

where K is a constant. If the changes in intensity and distance become infinitesimally small, the Δ becomes a differential, in the language of calculus. Then

$$\frac{dI}{I} = -K \, dx \qquad (5)$$

In order to relate I to x, this differential equation is integrated. Using the rules of calculus,

$$I = K_1 \exp(-Kx) \qquad (6)$$

where K_1 is another constant.

In Figure 6.2, the light intensity just below the surface of the water was the original light intensity minus the fraction R_s reflected, or $1-R_s$. Substituting this into Equation 6 with the value $x = 0$ representing the water surface,

$$I = 1 - R_s = K_1 e^0 = K_1 \qquad (7)$$

Putting this result back into Equation 6,

$$I = (1 - R_s) \exp(-Kx) \qquad (8)$$

In Figure 6.2, the light travels a distance D through the water before reaching the underwater foliage. Using Equation 8, the intensity at the leaf is

$$I = (1 - R_s) \exp(-KD) \qquad (9)$$

The leaf has a reflectance R_p so that Equation 9 is multiplied by this factor:

$$I = (1 - R_s) R_p \exp(-KD) \qquad (10)$$

The light leaving the leaf then travels up to the surface. The distance traveled may be slightly different from D, depending on the angles involved. For simplicity, the distance is assumed to be D. The intensity of Equation 10 is multiplied by a factor $\exp(-KD)$, obtaining

$$I = (1 - R_s) R_p \exp(-2KD) \qquad (11)$$

The absorption constant K remains the same, since the measurements are all in the same water. As the light again crosses the air-water surface, reflection again occurs. The fraction of light reflected is R_I. Multiplying Equation 9 by this term yields

$$I = (1 - R_s)(1 - R_I) R_p \exp(-2KD) \qquad (12)$$

Suppose the distance D, the reflectance R_s, R_I and R_p, as well as the intensity of light entering and leaving the water are known. K, the absorption constant, can then be calculated. When K is low, there is little absorption of light in the water, and the pollutants which cause water darkening are probably in low concentrations. When K is high, there is high absorption and probably higher levels of these pollutants. The value of K serves as a measure of this type of water pollution.

Simple theories can be combined in a chain effect to produce a relatively sophisticated method for detecting pollution. Its worth will have to be tested in the future.

MEASUREMENT

Chlorophyll and Optics

The level of chlorophyll in seawater can indicate the extent to which it has been polluted. The level can increase or decrease, depending on the type of pollutant, in contrast to most indicators of environmental degradation. For example, an area of the sea may naturally contain surface plants. If noxious sewage comes floating by, the plants may die off with a corresponding decrease in the chlorophyll level. Conversely, if plant food in the form of fertilizers are discharged into water, the plant population will explode. An algal "bloom" of this type almost choked Lake Erie to death. The presence of water plants can be a very sensitive indicator of the amount of pollutants in water.

132 PHYSICS OF THE ENVIRONMENT

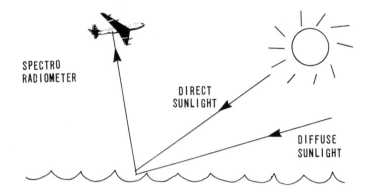

Figure 6.3 Measuring the level of chlorophyll on seawater by finding the light intensity changes. Seawater and chlorophyll absorb sunlight in different ways. An aircraft can detect chlorophyll levels over a wide expanse of water in a short time.[4]

A diagram of the system which can be used to find chlorophyll levels in water is shown in Figure 6.3. Sunlight, both from the sun itself and the sky, strikes the surface and is reflected upwards to an overhead aircraft. The measuring instrument on board is called a differential radiometer.

The experimental results are shown in Figure 6.4. The top curve is the absorption spectrum of typical water plants, like phytoplankton.

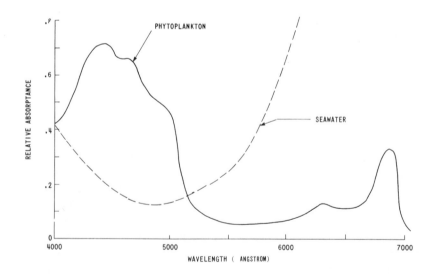

Figure 6.4 Change of absorptance for water plants and seawater with wavelength. Phytoplankton is a typical water plant in terms of absorptance spectrum. Its maximum absorptance is around 4,500 Å; the minimum for seawater is around 5,000 Å.[4]

LIGHT AND OPTICS 133

There is a maximum around 4,500 Å. The absorptance decreases rapidly with wavelength, but shows a maximum around 6,900 Å. The wavelength 4,500 Å is in the blue-green region, accounting for the greenish color of plants.

The dotted curve is the spectrum of sunlight reflected off seawater, with a minimum at around 5,000 Å. Seawater appears bluish-black due not to this minimum, but to the depth of water.

To discover the amount of chlorophyll on the water, absorptance at different wavelengths is compared. In practice, two wavelengths are used, at 4,400 and 5,300 Å, where the differences between phytoplankton and seawater are great and small, respectively.

The system devised for chlorophyll is based on finding a relatively small effect (water plants) in a big background (the entire ocean). This is a common problem in physics.

Figure 6.5 shows a set of experimental checks: measurements of the chlorophyll level along the American west coast. Using this graph, the signal from the radiometer can be correlated to the actual chlorophyll level. The radiometer converts the difference in absorptance of Figure 6.4 to voltage for simplicity in measurement.

The distances listed after the ocean data in Figure 6.5 indicate the approximate distance from shore. The spikes extending from points such as that for Lake Tahoe denote the standard deviations of the measurements, or the errors. As mentioned in Chapter 1, if a physical measurement is repeated a number of times, the answers usually differ. The spread of these answers is measured by the standard deviation. For Lake Tahoe, about two-thirds of all readings fell between -1.8 volts and -1.45 volts (the ends of the two horizontal bars). Similarly, the average chlorophyll level for that lake was about 1.7 milligrams per cubic meter, but about two-thirds of the readings fell between 1.2 and 3.

Ocean chlorophyll levels are not necessarily greater or lower than in a lake. The level depends on the type of lake. As well, there is a fairly strong decrease in chlorophyll level with distance out into the ocean, although the correlation is not perfect. Finally, there is a reasonable correlation between the levels measured by hand and the radiometer voltages found by remote sensing. This indicates that aircraft sensing is an accurate method of determining the concentration of chlorophyll in waters.

Figure 6.6 shows the chlorophyll variations in San Francisco Bay. There is a definite decrease between stations 3 and 6, encompassing the most heavily populated section of the Bay. This indicates that pollution from this area is probably killing normal plant life.

Ocean and lake temperatures depend on the weather. In turn, water plant life is highly sensitive to water temperature. One might think that

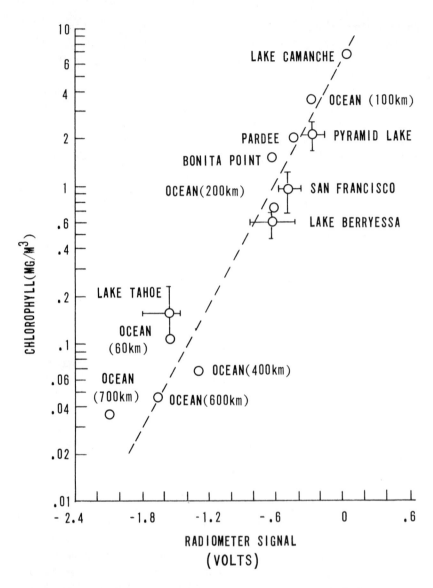

Figure 6.5 Comparison of the biological with the airborne method of measuring chlorophyll levels. The measurements were taken at locations on the west coast of the United States. The points fall approximately on a straight line on the semilogarithmic set of scales. Lengths noted by the ocean points are the distances from the shore. Symbols are explained in the text.[4]

LIGHT AND OPTICS 135

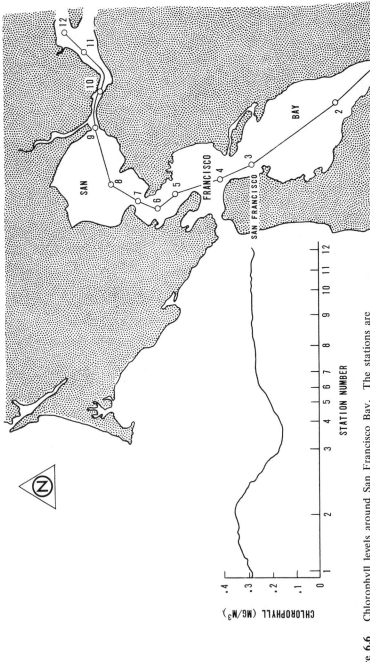

Figure 6.6 Chlorophyll levels around San Francisco Bay. The stations are only arbitrary points on the water. The chlorophyll level declines sharply between stations 2 and 6. Enough readings of this type will indicate the eutrophication state of this and other bays and lakes.[4]

the warmer the water, the greater the plant growth. However, this is not always the case, as is shown in Figure 6.7, where water temperature and chlorophyll level were measured simultaneously.

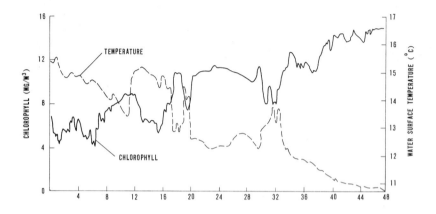

Figure 6.7 Change of chlorophyll level and seawater surface temperature with distance from shore in km. When the temperature rises, the chlorophyll level decreases, with few exceptions.[4]

Figure 6.7 shows an approximately inverse relationship between chlorophyll levels and temperature. This may not hold for every area, since the levels are often governed by the concentration of plant nutrients. However, knowledge of the change of temperature level is useful. Knowing how the growth of water plants is affected by heat from thermal power stations may prevent the overgrowth of a lake.

The absorptance properties of water plants and seawater may be used to distinguish between the two. Information on wide expanses of vegetation growth can then be found. This growth, or lack of it, is frequently a graphic indication of pollution washing down from the land.

Oil and Light

The radiometer is used to measure quantities other than chlorophyll. Figure 6.8 shows the radiance ratio of oil-covered water to that of normal water. The radiance of oil is usually higher than that of seawater.

On clear days, there is almost no variation between the two; Figure 6.8 shows that on partly overcast and totally overcast days there is at least one maximum, indicating the presence of an oil slick. This radiometric method can then be used to find these slicks.

LIGHT AND OPTICS 137

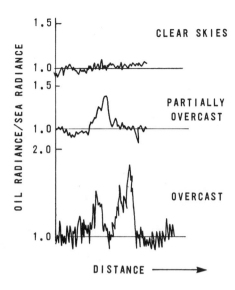

Figure 6.8 Change of the ratio of radiance of oil-covered water to seawater with degree of cloudiness. The greater the degree of cloudiness, the easier it is to see oil slicks. The x-axis denotes distance along the water surface.[5]

Since Figure 6.8 is raw experimental data, the curves are more jagged than usual. This irregularity is due to the extreme sensitivity of the measuring instrument.

Figure 6.9 shows hypothetical experimental results. They appear to follow a linear pattern, at a slope of about one-third. One way to smooth out the points is to average the two adjacent points and plot this in the middle, instead of the two original points. For example, suppose the experimental points for a certain measurement, at equal intervals starting from zero, were 0, 1.5, 2.5, 2.0, 2.5, 2, 1, 0 and -2. Using the above procedure for averaging would yield 0.75, 2, 2.25, 2.25, 2.25, 1.5, 0.5 and -1.

Because finding oil on the sea is dependent on so many variables, including weather, wave height and the angle at which the light is measured, these variables must be evaluated carefully. By controlling one particular variable, oil slicks on the ocean can be found.

Oil Identification

Oil spillers are notoriously reticent. Somehow these offenders must be identified and controlled.

138 PHYSICS OF THE ENVIRONMENT

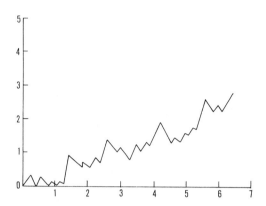

Figure 6.9 Hypothetical example of variation of one quantity with another. Although this graph is imaginary, many experiments show this jagged behavior.

First, the spilled oil must be identified. Then, after learning the movement of the ships in that area and their cargo, the culprit can be uncovered. The problem, however, is identifying the oil.

While oil from Texas looks exactly like oil from the Middle East, different oils absorb infrared light in different ways. An oil could be identified as offshore Louisiana, for example, if it absorbed a great deal of light at a wavelength of 85,500 Å.

This absorption method was used in the Santa Barbara oil blowout in the early 1970s. It became necessary to differentiate between oil that was being blown out of Platform A, the source of most of the spillage, and the natural oil seepage which occurs in the Santa Barbara channel. Using infrared absorption, scientists could tell the difference between the two types.

There are hundreds of oil fields. Table 6.2 lists 20 crude oils from all over the world. Each oil shown in this table absorbs light in a slightly different way because of its varying molecular structure. When the absorption vs wavelength is plotted, a curve similar to Figure 6.10 is obtained. Another oil might have a slightly different curve. How can they be told apart mathematically?

One way is to compare the areas of the well-defined peaks—those that do not overlap. For example, in Figure 6.10 peaks D and E would not be used, since they cross in the center. Peak A's area is well-defined and suitable for measurement since it is relatively solitary. That the area of peak A is x units and the area of another oil's absorption peak at the

LIGHT AND OPTICS 139

Table 6.2 Crude Oils Used in an Infrared Absorption Experiment[6]

Number	Origin
1	East Texas Field
2	El Morgan, Egypt
3	Wilmington, California
4	Minas Field, Sumatra
5	Duri Field, Sumatra
6	Export, Neutral Zone, Saudi Arabia-Kuwait
7	Kuwait Export
8	Ceuta, East Shore Lake Maracaibo, Venezuela
9	Mesa, Orinoco Basin, Venezuela
10	Timbalier Bay, Offshore Louisiana
11	Sarir, Sirte Basin, Libya
12	Agha Jari, Iran
13	Orita, Putumayo Basin, Columbia
14	Kuwait Export Blend
15	Arabian Light, Export Blend
16	Ward-Estes Field, Texas
17	Goldsmith Field, Texas
18	Kelly Snyder Field, Texas
19	Sprayberry (Trend Area), Texas
20	Headlee, Texas

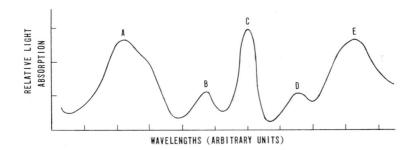

Figure 6.10 Changes of relative light absorption with wavelength. The graph is made up of peaks which sometimes run into each other.

same wavelength is y units suggests a simple method of comparing the two oils. The wavelength of A must duplicate the wavelength of A′, where A′ is the analogous peak for another oil.

Table 6.3 shows eight peaks that fall at almost exactly the same wavelength from one oil to the other. There are small differences because of molecular structure and chemical composition. Although the

Table 6.3 Characteristics of Eight Peaks Chosen to "Fingerprint" Oil[6]

Peak	Wavelength (Å)	Wavelength Range	Standard Deviation of Wavelength (Å)	Range of Peak Areas (arbitrary units)
1	59,000	58,300- 59,500	520	0.0175-0.0920
2	62,500	62,200- 62,700	260	0.0370-0.4140
3	85,600	84,200- 86,600	1000	0.0074-0.0311
4	95,800	95,800- 97,000	520	0.0068-0.1896
5	114,000	112,000-116,000	1400	0.0220-0.1509
6	123,000	122,000-123,500	780	0.0076-0.0920
7	134,000	132,000-135,000	1000	0.0013-0.0546
8	138,000	136,000-139,000	1000	0.0011-0.0922

range of the wavelengths for each peak is shown, the most important figure is the standard deviation—the range into which about two thirds of the wavelengths fall. For the first peak, about two thirds of all the oils measured fall within the range 58,500 ($\cong$ 59,000-520) to 59,500 ($\cong$ 59,000 + 520) Å.

In the last column, the ranges of the peak areas are shown. Each of the eight area ranges was itself subdivided into eight "blocks." For example, for peak 1 the first two blocks were from 0.0175 to 0.0268 [= (0.0920 - 0.0175)/8 + 0.0175] and 0.0268 to 0.0361. Suppose the first peak of a particular oil had an area of 0.03 units. It would then fall into the second block, and the first digit of its eight-digit fingerprint would be 2. If the area of its second peak was 0.18 units, its second digit would be 4.

This process is called digitizing—transforming ordinary numbers and letters into groups of digits. For example, the letters of Jones could be written as 1015140519, where each letter has been digitized by giving it its numerical rank in the alphabet: "J" is the 10th, "O" is the 15th and so on.

To get an eight digit "fingerprint" for each of the 20 oils, the area of each peak is measured to find which block it falls into. Results of the experiments are shown in Table 6.4. Each oil has a unique fingerprint.

A relatively simple experiment can differentiate complex materials. With it oil spillers can be positively identified.

Quantifying Smell

The more obvious **types** of pollution are those occuring in air and water. Other forms of pollution—odor pollution, for example, are less

Table 6.4 "Fingerprints"[a] of the 20 Oil Samples in Table 6.2[6]

Sample	Identifying Numbers
1	72622341
2	53363412
3	53442351
4	22311124
5	75115563
6	22563541
7	33563621
8	23454531
9	63664521
10	62843431
11	42282433
12	12242421
13	33533431
14	32863522
15	11543522
16	22853421
17	21832321
18	32822321
19	21511211
20	11311115

[a]These are a series of identifying numbers, like a telephone number, each representing the relative value of the area of an infrared absorption peak.

publicized simply because they are so difficult to measure. How can a number be put on smell?

Chapter 3 showed that noise has both a physical and psychological effect. The intensity and frequency of sound can be measured, but its degree of irritation depends on psychological factors such as duration and time and frequency of occurrence. Similarly, the presence of certain chemicals in the air can be measured. Why some are more irritating than others is generally unknown. If more was known about reactions to smell, scientific standards and laws could be formulated to control industrial odors.

The sound of a jackhammer doesn't seem as irritating after a few hours as it did the first few seconds. The object of the following experiment was to analyze this "familiarity" effect.

Methylmercaptan, the prime component of many bad smells, was chosen for the experiment. Air samples containing methylmercaptan were passed through a hydrogen flame, which has a simple spectrum. When sulfur, a constituent of methylmercaptan, is heated in the flame, a strong violet light is produced, corresponding to a particular frequency in its spectrum.

142 PHYSICS OF THE ENVIRONMENT

The experimental setup is shown in Figure 6.11. Hydrogen and the polluted air enter a narrow tube at the bottom. A flame is lit at the center of the apparatus. The emitted light from the flame is passed through a filter, which eliminates all wavelengths of light except the violet. The light is then "amplified" by a photomultiplier tube, which produces a voltage proportional to the intensity of light it detects.

Physical results are shown in Figure 6.12A. There are definite maxima around 20 seconds, 55 seconds and 255 seconds. The maxima illustrate the great variability in spite of methylmercaptan being sprayed into the air at a constant rate.

Human results are depicted in Figure 6.12B. Observers were stationed near the measuring instruments, each with a set of five pushbuttons. The first indicated a perceived faint smell, the second a somewhat stronger smell, and so on up to number 5 for a strong smell. Only the first three pushbuttons were actually used, indicating that the perceived smells were not particularly strong.

In the first minute of the experiment, the concentration increases rapidly around 15 seconds, and so does the response. At around 25 seconds, the concentration drops with the response. Around 33 seconds, the response is 2, with no measured concentration at all. A similar effect occurs at 38 seconds. The observers note the increase in concentration after 40 seconds, but completely miss the peak around 53 seconds.

In order to eliminate objectionable odors, human reactions must be known, and the experiment shows the difficulties. This subject is at the stage that the science of human response to sound was 50 or 100 years ago.

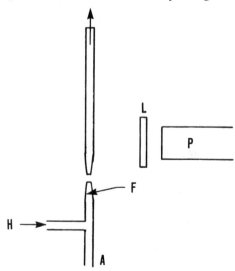

Figure 6.11 Detection of odor in air. The polluted air is drawn into the instrument through A. Hydrogen is added at point H, and the mixture is lit. Flame F gives off a spectrum of light, which passes through filter L, so the instrument which measures the light detects only particular wavelengths. The photomultiplier P "amplifies" the light passing through the filter. Its change of intensity is recorded as a change of voltage.[7]

LIGHT AND OPTICS 143

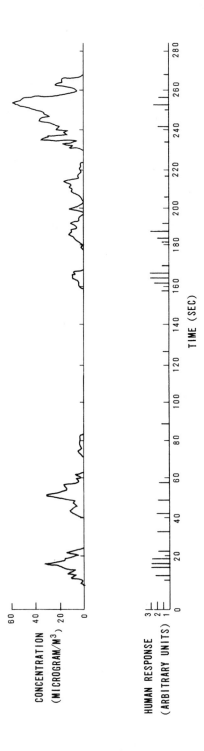

Figure 6.12 Concentration of methylmercaptan. Part A shows the **results** of measurements by the method described in Figure 6.11. Although the chemical was released at a constant rate, winds changed the measured concentration to a series of maxima and **smell**-free periods. The corresponding human response, as shown in part B, is measured by a series of pushbuttons.[7]

Remote Detection

The cameras mounted in unmanned satellites can see mountains, oceans and weather patterns. If they are sophisticated enough, they can sense some types of air pollution, such as dust and smoke billowing from the factories. Polluted areas are detected by noting color differences between these areas and the surrounding countryside.

The same principles hold when the earth is photographed from an aircraft. How is the photo interpreted in terms of air pollution travel and where it is concentrated?

The first step is to measure the darkness or lightness of the picture, using an instrument called a microdensitometer. This instrument indicates the exact variation in darkness along each of the lines. Results of a flight near Tacoma, Washington are shown in Figure 6.13. If a portion is completely white, it has a relative transmittance of 1; if it is completely black, the relative transmittance is 0.

Degrees of picture whiteness are translated into a measure of air pollution. The whiter the area, the more pollution between the ground and the overhead aircraft. There are exceptions to this rule, however, which depend on the color of the pollutants.

What is the relationship between transmittance and air pollution? To answer this, recall the Beer-Lambert law. Rewriting Equation 6, the intensity of light passing through air or water decreases with the distance through which it passes, by the formula

$$I = K_1 \exp(-Kx) \qquad (13)$$

where I = the intensity
 K_1 & K = constants
 x = the distance the light has travelled

If air pollution increases, the area it covers in the photograph becomes whiter, and the relative transmittance increases. Using the same principles as the Beer-Lambert law, the transmittance can be written:

$$T = K_3 \exp(-K_2 x) \qquad (14)$$

where K_3 is yet another constant.

K_2, the absorption constant, is a measure of the degree of air pollution. It is now found using Equation 14. K_2 is an indication of how fast light is being absorbed by the atmosphere at any point. Since many air pollutants absorb light, this quantity is a measure of air quality, or lack of it.

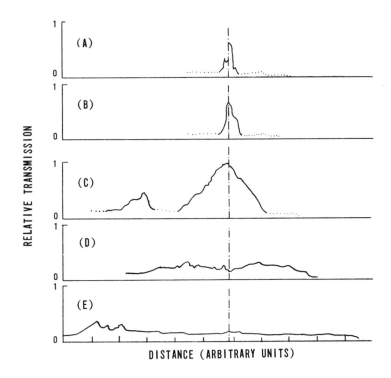

Figure 6.13 Relative transmission along each of five cross sections of an aerial photograph taken near Tacoma, Washington. The central section of line C is the whitest on the photograph, and this is demonstrated here. Similarly, by the time the air pollution blows over to lines D and E, it has been dispersed widely by the wind. As a result, there is little effect in the last two graphs.[8]

Earth satellite data can be used to monitor the environment. Aircraft photographs and mathematics can be used to get a number characterizing the degree of air pollution at each point.

Sulfites and Sulfides

Sulfur air pollutants may appear as at least three types of ions: sulfate (SO_4), sulfite (SO_3) and sulfide (S_2). Sources of atmospheric pollution often give off different combinations of compounds, emitted at different times of day and under different weather conditions. In the case of smog, for example, the amount of sulfate and sulfite ions in air pollution varies according to the hour, **the** types of fuels being burned and the sun's reaction with the ions. Finding the types of ions in air pollution tells what is coming out where, and how the weather is affecting it.

The phenomenon used is called the photoelectric effect. Photons (or light) are beamed onto the air particulates which have been collected on an air filter. If the light has the right wavelength, it interacts with an electron in one of the pollutant atoms. The electron is expelled from the atom, and its energy can be measured. The process is shown in Figure 6.14. The descriptive equation is

$$E_k = E_p - E_b \qquad (15)$$

where E_k = the kinetic energy of the electron which leaves the atom when struck
E_p = the energy of the photon
E_b = the "binding energy" of the electron (the energy it takes to set it free)

The equation for the energy of the photon is

$$E_p = hf \qquad (16)$$

where h is Planck's constant, and f is the frequency of the light or photon.

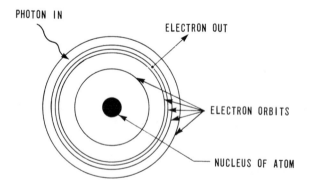

Figure 6.14 Schematic representation of the photoelectric effect. A photon, on the left, interacts with an electron in orbit. The electron then leaves the atom, on the right. The expelled electron's kinetic energy depends on the frequency of the incoming photon and the orbit occupied.[9]

Although there are a number of ways to measure the energy of electrons after they have left the atom, one of the simplest is shown in Figure 6.15. When a magnetic field is applied to a moving charged body like an electron, its path curves. The curvature radius will depend on its energy. If the energy is high, the radius is large; if the energy is low, the radius is small. The kinetic energy E_k of the electron can then be found.

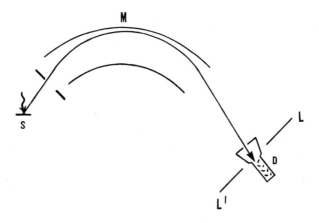

Figure 6.15 Finding the kinetic energy of a photoelectron. Light strikes the sample S and an electron is released into the magnetic field M. Its radius of curvature is determined by its original kinetic energy. The detector D is moved along the line LL$'$, so that the number of electrons is found as a function of their radius of curvature.[9]

The binding energy varies with the chemical structure of which the sulfur atom is **part.** Because there is an extra oxygen atom in the sulfate ion, the binding energy of the electrons coming from it, as found from Equation 15, will be different from those of the sulfite ion. These differences are used to determine types of air pollution.

An example of photoelectron spectroscopy, as the process is called, is shown in Figure 6.16. The maxima go back and forth between two lines labelled I and II, representing the sulfate and the sulfite ion, respectively. Depending on the time of day in Pasadena, California, where the measurements were taken, one of the two ions is prevalent.

148 PHYSICS OF THE ENVIRONMENT

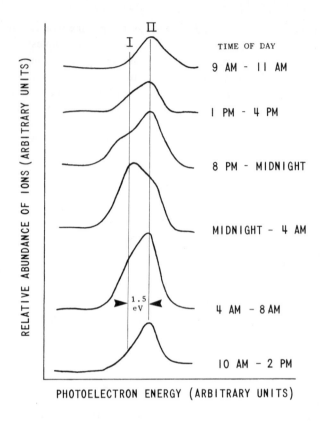

Figure 6.16 Variation in the relative abundance of sulfate (peak at I) and sulfite (peak at II) ions in smog. Arbitrary units are used for the x- and y-axes of all curves. The maximum abundance of the sulfate ions occurs late at night.[9]

For most of the day, the sulfite ion has a greater relative abundance. However, its ratio to sulfate ions decreases with nightfall, and by midnight the latter ion has a higher concentration.

By using light and the photoelectric properties of matter, chemical states of the same element can be differentiated. These states vary considerably in some forms of air pollution, like smog. A full understanding of pollution requires knowledge of how sulfates and sulfites, nitrates and nitrites, and all the other ions change in abundance with time and weather conditions.

CONTROL

Ultraviolet Light in Water

Because of wastes dumped into water, eating shellfish can cause a viral epidemic. The first step in halting these outbreaks is to stop the pollution causing them. While adding chlorine to water kills most of the bacteria in it, shellfish can suffer when chlorinated water is passed through them, so other methods should be tried.

Exposing contaminated water to ultraviolet (UV) light kills most of the bacteria and viruses it contains, because the light energy destroys parts of the cells it enters. Although bacteria could be eliminated by employing heat or electricity, the use of light requires the least energy.

Table 6.5 shows the results of shining UV light on contaminated seawater. The first group deals with static or nonflowing seawater, and the second is concerned with flowing water. The UV light kills most of the viruses in a short time.

Table 6.5 Eliminating Viruses by Ultraviolet Light[10]

Virus Type	Initial Virus Count	Virus Count after 15 sec	Virus Count after 30 sec
Viruses in Still Water:			
Poliovirus 1	8,074	78.7	2.9
Poliovirus 2	4,746	64.2	3.2
Poliovirus 3	49,140	669.9	16.4
Echovirus 1	15,603	142.2	5.7
Echovirus 11	7,872	132.4	5.7
Coxsackievirus A-9	65,200	896	58.4
Coxsackievirus B-1	88,000	3,072	424
Reovirus 1	4,800	480	48
Viruses in Moving Water:			
Poliovirus 1	880	0.2	less than 0.2

The water may then be discharged, providing that other pollutants have been removed as well. The application of UV light to water can remove most of the harmful life in it. Since shellfish have to use that water eventually, the use of optics can help.

Pesticides and Light

Pesticides have received little attention here, because so much of the work in tracing their path in food is chemical. However, physics can play a part in eliminating pesticides from one food—milk.

Pesticides get into milk when sprayed crops are fed to cattle. They can be removed by chemical and other methods, but this can be expensive and may damage the milk.

Some pesticides are photosensitive, *i.e.*, they undergo a reaction when exposed to the right wavelength and intensity of light, just as photographic film does. When exposed to UV light, they break down into simpler chemicals which are not harmful to humans.

Figure 6.17 shows the magnitude of the photosensitivity. The effect of pesticide destruction varies considerably with the type involved. Methoxychlor is destroyed fastest, heptachlor epoxide slowest.

SUMMARY

In this chapter, many facets of using light to solve environmental problems have been explored, ranging from detecting oil to eliminating pesticides. While all the problems have not been solved, optics can make a large contribution.

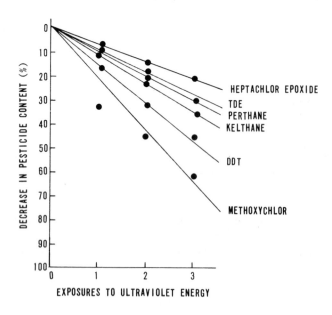

Figure 6.17 Destructive effect of UV light on six pesticides in milk.[11]

PROBLEMS

6.1 Do the graphs of Figure 6.1 provide enough uniformity to allow the detection of asbestos, which could be made up of any or all of the three types, in factory air? Explain.

6.2 The graphs of Figure 6.1 can be combined into one "average asbestos" graph. Average the values of transmittance at 3, 5, 7, ... 13 x 10^4 Å, and plot this graph. Is the detail of the three individual graphs smudged out? Is this average suitable for the detection of asbestos?

6.3 There might be other ways to characterize each of the three graphs in Figure 6.1. For example, find the position of the minimum of transmittance for each. Would this single number be adequate to characterize them and their average? Would the positions of the maxima be better or worse? Why?

6.4 Plot the size of a bank account (at a 5% compound interest rate) for 10 years on linear graph paper. Is it a straight line? What would the shape of the Beer-Lambert law be on this graph?

6.5 If R_s is 0.2 and K is 2 cm^{-1}, at what distance does the light intensity in Equation 8 decrease to 0.5? to 0.2?

6.6 The values of R_s and R_p are important to the underwater foliage method of detecting water pollution. For example, if $R_p = 0$ for a given wavelength of light, an experiment at that wavelength would not work. Suppose Figure 6.1A showed the reflectance with changing wavelength, rather than the transmittance. Discuss which wavelength would be best for the present experiment, and which should be avoided.

6.7 Assuming that no light is absorbed by the leaf in the underwater foliage experiment, what is the light intensity transmitted by it?

6.8 Let the three terms in front of the exponent of Equation 12 combine into a value of 0.7. If D = 2 m, what is the value of K if the intensity of light coming out at the surface is measured to be 0.5? 0.2?

6.9 List some assumptions which went into the derivation of Equation 12.

6.10 What does the Beer-Lambert law have to do with the "black box" effect?

6.11 Plot the difference in absorptance for seawater and water plants from the wavelength 4,000 to 5,500 Å. Is the choice of 4,400 and 5,300 Å a good selection for the two reference wavelengths?

6.12 On the basis of Figure 6.4, would a reference wavelength of 7,000 Å have been useful? Explain.

6.13 What is the equation of chlorophyll level vs radiometer signal in Figure 6.5?

6.14 On the basis of Figure 6.5, what are the approximate highest and lowest chlorophyll levels which can be detected? If a change in the radiometer signal of 0.1 unit can be detected, what is the smallest change in chlorophyll level that can be found?

6.15 For the points in Figure 6.5 listing ocean chlorophyll levels, plot the chlorophyll level as the distance from the shore varies. Is there a correlation? Why? What accounts for the one apparently misplaced point? In view of this, is a further experiment necessary to prove the correlation of distance and chlorophyll level?

6.16 What is the largest spread in chlorophyll levels in Figure 6.5? What is the largest spread in radiometer voltages?

6.17 From Figure 6.5, Lake Camanche has a high level of chlorophyll. What is the ratio of its level to that of Lake Tahoe?

6.18 Lake Camanche is similar to Lake Erie in that it is eutrophic—containing a high amount of water plants. What type of lake is Lake Tahoe? A biology text may be checked to find the answer. List possible reasons, both natural and manmade, for the differences among the lakes in Figure 6.5.

6.19 How would tides affect the results of Figure 6.6?

6.20 In Figure 6.6, near which station is the greatest variation of level with distance? Near which station, or stations, is this variation lowest?

6.21 Figure 6.7 seems to show that when the water temperature is lower, the level of chlorophyll rises. Is this really true? Plot the inverse of the chlorophyll level with distance for every 2 km from 20 km to 48 km. Then on the same graph plot the temperature. Use right- and left-hand scales, as in Figure 6.7. Explain the conclusions. Would any correlation hold true for even lower temperatures?

6.22 What would be the likely effect of thermal power stations on chlorophyll levels, on the basis of Figure 6.7?

6.23 Using Figure 6.8, describe a method for characterizing the ability to detect an oil slick.

6.24 Plot both the first and second set of points in the example used in the discussion of Figure 6.9. Which curve is smoother? Would the result have been smoother if points had been taken three at a time? Is any information lost in this smoothing process? Explain.

6.25 Follow the two-by-two procedure and smooth out the graph of Figure 6.9. Use intervals of about 0.2 units along the x-axis. Does it resemble a straight line now?

6.26 Suppose the average sea radiances for the clear, partly overcast and completely overcast days in Figure 6.8 are 10, 2 and 0.5 units. Replot the three graphs on one piece of graph paper. In effect, Figure 6.8 has been denormalized. Is the graph easier to read?

6.27 Which peak in Table 6.3 has the lowest relative standard deviation? It can be defined as S/W, where S is the standard deviation of the peak and W is its wavelength. The relative standard deviation is another measure of scattering and takes account of the size of the quantity as well as its standard deviation.

6.28 In Table 6.3, is there a relationship between the wavelength and the standard deviation of the peaks? Explain.

6.29 In Table 6.3, what would be the digit for a seventh peak with an area of 0.04 units?

6.30 How many fingerprints in Table 6.4 have 6 digits in common? Oil from the same area of the world is more likely to have similar fingerprints. Would this degree of coincidence pose a problem for the method?

6.31 How many fingerprints are possible with the technique shown in Table 6.4?

6.32 If each number in an eight-digit oil fingerprint were chosen at random, what would be the chances of duplication among the 20 cases?

6.33 What problems are there in maintaining the accuracy and reliability of the oil fingerprint method? For example, do oils always have the same absorption spectrum no matter how long they remain on the ocean?

6.34 In Table 6.3, which peak has the greatest spread in area? Which has the greatest ratio of its highest value of area to its lowest value? Which basis is best for choosing the peaks to incorporate into the eight?

6.35 Sign your first name in big letters on graph paper. Divide the y-axis into eight sections, letting the height of the tallest letter be in the eighth block. Write out the "fingerprint" by determining into which block each of the peaks in your name falls.

6.36 Using the maximum concentration in Figure 6.12A as equivalent to a response of 5, plot the corresponding expected responses for the major maxima of concentration. Do the actual responses generally correspond to these expected responses? How many false alarms occurred in the five minutes of the experiment?

6.37 Based on the first—up to about 60 seconds—and second—after about 160 seconds—sets of maxima of concentrations in Figure 6.12A, is the observer's nose getting jaded? Explain. This point could play a large role in explaining the often contradictory results in human reaction to smell.

6.38 Are there time lags in the responses to maxima in concentration in Figure 6.12? Give some examples, and estimate the lags. One case is the response at about 90 sec. If there were always a time lag, how could it be accounted for in designing an experiment? Would the possibility of getting accustomed to smells complicate the problem of looking for time lags? Try to explain this in a mathematical way, drawing graphs if necessary.

6.39 Would the microdensitometer method be applicable to pollution from a coal-burning power station? Investigate the color of different air pollutants to see which could be detected using this method.

6.40 Justify more fully, on the basis of physical considerations, the choice of Equation 14 to relate T and the absorption constant of air pollution K_2. Would a change in the type of air pollution also change K_2?

6.41 Assume that $K_3 = 1$ and $x = 3 \times 10^3$ m. Then calculate K_2 at the ten distance marks on parts A and E of Figure 6.13.

6.42 What is the maximum and minimum value of K_2 for each curve of Figure 6.13? What is the ratio of this maximum and minimum? Which curve has the lowest ratio?

6.43 In the discussion of Figure 6.13, it was stated that the air pollution became diffuse by the last two lines. Using the maximum values of K_2 for the three graphs calculated in problem 6.41, is this true?

6.44 At what point in Figure 6.13 is the rate of change of K_2 with distance greatest? What does this probably indicate?

6.45 Could there ever be an electron with a binding energy of zero? Discuss what this would mean physically.

6.46 Take the lowest level of each curve in Figure 6.16 as a relative abundance of zero, and the maximum of each as an abundance of 1. Then plot the ratio of the abundance of sulfate ions to that of sulfite ions with time, from 10 AM to noon of the next day. Take the midpoint of the times given for each curve as the x-point on the new graph. For example, the first curve would have a relative abundance of about 0.2 and a time of 10 AM, halfway between 9 AM and 11 AM. At what time would the greatest relative abundance of sulfate ions occur? What is the corresponding time for sulfite ions?

6.47 What is the significance of the times of greatest relative abundance in Figure 6.16? Hint: do chemical compounds change with time?

6.48 Plot the virus counts as time varies for the first five viruses of Table 6.5, on both regular and semilogarithmic graph paper. Which produces the better straight line through each set of three points? What would be the general form of equation for the virus counts vs time on the semilogarithmic plot? Using this type of paper, estimate the virus count for reovirus 1 after 1 min.

6.49 In Table 6.5, does having the seawater flow past the UV light kill the bacteria faster than having the water stationary? Why does this occur?

6.50 Suppose it were required by law that 99.9% of any type of virus be killed before sewage water was declared safe for discharge into oceans and lakes. How long would the water have to be treated with UV light to achieve this goal for the viruses listed in Table 6.5?

6.51 Would it be worthwhile changing the frequency of the UV light in Table 6.5 to see if its effect changes? Hint: What is the effect of ordinary sunlight on viruses?

6.52 Find out independently how far UV light penetrates into water. How would this limitation affect the method of eliminating viruses?

6.53 Find the equation of each of the lines of Figure 6.17 letting the percentage remaining be y and the number of exposures be x. How many exposures would it take to decrease the heptachlor epoxide content by 95%?

6.54 Will the straight lines in Figure 6.17 yield a 100% decrease? Give a tentative physical reason.

6.55 Could the lines for each pesticide in Figure 6.17 be characterized by one number? What is it?

6.56 Suppose a given quantity of milk contained 2, 4, 6, 8 and 10 units, respectively, of heptachlor epoxide, TDE, kelthane, DDT and methoxychlor. Plot their decrease with the number of exposures. Which pesticide first goes below the level of 0.1 unit? How many exposures does it take?

REFERENCES

1. *Aerial and Orbital Remote Sensing of Water Quality.* Space Division, North American Rockwell Corporation, Report SD71-471 (February 1972), p. 5.
2. Noonan, F. M., and B. Linsky. "Internal Reflection Spectroscopy Applied to Air Pollution," *Atmos. Environ.* 4(2):126-128 (1970).
3. Spooner, D. L., and L. R. Lankes. "Determination of Water Depth by Remote Measurement of Reflectance of Underwater Plants in the Near-Infrared," presented at American Society of Photogrammetry, Washington D.C., March 1, 1970.
4. Arvesen, J. C., J. P. Millard and E. C. Weaver. "Remote Sensing of Chlorophyll and Temperature in Marine and Fresh Waters," presented at 23rd International Aeronautical Congress, Brussels, Belgium, September 1971.
5. Millard, J. P., and J. C. Arvesen. "Airborne Optical Detection of Oil on Water," *Appl. Opt.* 11(1):105 (1972).
6. Mattson, J. S. " 'Fingerprinting' of Oil by Infrared Spectroscopy," *Anal. Chem.* 43(13):1872, 1873 (1971).
7. Barynin, J. "Measuring Odour Pollution," *New Scientist* 48(723):118 (1970).
8. Veress, S. A. "Air Pollution Research," *Photogram. Eng.* (August 1970), p. 847.
9. Novakov, T. "Photoelectron Spectroscopy for Identification of Chemical States," presented at Joint Conference on Sensing of Environmental Pollutants, Palo Alto, California, November 8, 1971, p. 4.
10. Hill, W. F., Jr., E. W. Akin, W. H. Benton and F. E. Hamblet. "Viral Disinfection of Estuarine Water by UV," *Proceedings of the American Society of Civil Engineers, Journal of the Sanitary Engineering Division,* SA5 (October 1971), p. 605.
11. Li, C. F., and R. L. Bradley, Jr. "Degradation of Chlorinated Hydrocarbon Pesticides and Butteroil by Ultraviolet Energy," *J. Dairy Sci.* 52(1):28 (1969).

CHAPTER 7

LASERS

"The laser: a solution in search of a problem."
Physics proverb.

Goldfinger: Mr. Bond, this won't hurt a bit.
From *Goldfinger*,
released by United Artists

INTRODUCTION

Since many physics texts give little mention to lasers because of the advanced physics involved, a brief review of the physics of lasers will follow.

Laser Operation

What makes light from a laser different from that of a flashlight? The main differences are in intensity, directionality and coherence. Consider directionality first. The laser beam is extremely directional compared to ordinary light. It spreads out, in long distances, only about five parts per million of the distance travelled. In 1 km the beam width at the end of its journey will be about 0.5 cm. An ordinary source of light will be completely diffuse by the time it travels a hundred meters or so. While it can be seen at that distance, its power is negligible because of the spread. All of the laser power—allowing for small losses in air—will be in that 0.5 cm diameter circle. If a mirror were held to it, the light could be reflected, without much more spreading, back to the source.

How is this directionality achieved? The key is the use of the mirror at the end of the laser beam's travels. Photons—short bursts or quanta of light—which do not move parallel to the length of the laser are eliminated.

158 PHYSICS OF THE ENVIRONMENT

In Figure 7.1, the large cylindrical object in the center is the "active ingredient." Made of a crystal, such as ruby (or sometimes a gas), it glows like a fluorescent tube when light falls on it. The coils in Figure 7.1 supply the light energy. A powerful source of light, the coils or flashtubes are similar to electronic flashes.

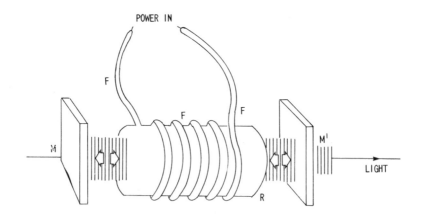

Figure 7.1 A ruby laser. The helical flashtube F encircles the cylindrical ruby crystal R. Mirrors M and M' (the latter is half-silvered) reflect the light back and forth through the crystal. The laser beam emerges through the right-hand mirror. In more modern devices, the ruby is often replaced by gases such as mixtures of helium and neon.[1]

Suppose a photon is created by the flashtube somewhere near the middle of the ruby, and travels toward the left. Its direction of travel is a few degrees from the axis of the ruby, as shown in Figure 7.2. When the photon strikes the mirror M on the left-hand side, it will be reflected by the laws of optics—angle of incidence equals angle of reflection—at an angle θ away from the ruby axis. This is indicated by the arrowhead pointing toward the right. By the time the photon gets back to the same horizontal position in the ruby that it had originally, it has descended a distance 2d tan θ, where d is the distance the photon travelled to reach the first mirror.

The photon continues to "bounce" back and forth between the two mirrors. Each time it does, it drops a distance 2L tan θ, where L is the ruby's total length. If the number of bounces or the angle θ is large, the photon strikes the bottom of the ruby and passes out of the laser. But if θ is zero or very small—i.e., the photon travels parallel or almost

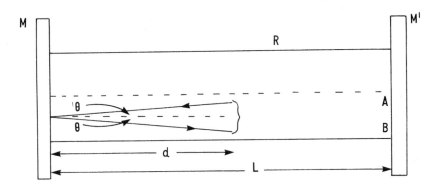

Figure 7.2 Travel of a photon in a laser when the photon moves at an angle θ from linearity with the active element R. M and M' are mirrors, as in Figure 7.1. The latter is half-silvered. The distance represented by the bracket is $2d \tan \theta$.

parallel to the ruby's axis—some of the photons reflecting back and forth will eventually pass out through the half-silvered mirror on the right-hand side. It may take anywhere from one to an almost infinite number of reflections. "Half-silvered" implies that the mirror reflects only part of the light on it and lets the rest through.

The "ser" in laser stands for *stimulated emission of radiation.* As the photon travels along, it tends to pick up other photons from the atoms it strikes. They move in the same direction and in the same phase as the original. There is thus a multiplying effect. Even if some of the original photons are absorbed, there are enough extra photons to keep the beam bright.

The elimination of photons whose direction is not parallel to the main axis and stimulated emission of radiation tend to make the beam emerging from the laser highly directional.

The intensity of a laser is great because the strongly directional light is confined to a small area. A typical laser can reach power densities of about 108 W/cm^2. If the power of a million ordinary light bulbs were focused on an area the size of a fingernail, the power density would be the same as that of a laser.

A laser beam is coherent. When an electron is an atom that falls from an excited high-energy state to a nonexcited lower-energy state, it emits a photon. A typical energy level diagram is shown in Figure 7.3. The frequency of the light emitted depends on the length of the "fall," and is governed by the equation

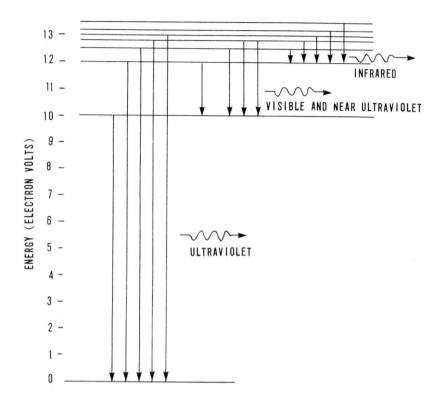

Figure 7.3 Energy levels of the hydrogen atom, with the energy scale in electron volts. When the atom changes from a higher state to a lower state, as indicated by the vertical arrows, it emits a packet of electromagnetic radiation whose frequency is proportional to the energy change involved, *i.e.,* to the length of the arrow. In actual lasers, the active material used has a more complicated energy level diagram than simple hydrogen.[1]

$$E_2 - E_1 = hf \qquad (1)$$

where E_2 and E_1 = the two energy levels between which the electron drops
 h = Planck's constant
 f = the frequency of the light

In an ordinary material, almost all the electrons will be unexcited. The excited ones make lasing possible.

For ordinary (nonlasing) gases and solids, the relative number of electrons at each level is governed by Boltzmann's ratio:

$$\frac{N_m}{N_n} = \exp\{-(E_m - E_n)/kT\} \quad (2)$$

where N_m = number of electrons in level m
       N_n = number of electrons in level n
       E_m & E_n = the energies at the two levels
       k = Boltzmann's constant
       T = absolute temperature

The higher the energy of a level, the fewer the electrons at that level. For simplicity, temperature effects are disregarded.

The equation holds for the normal state of affairs. Suppose there was a "population inversion"—an energy level with more electrons than levels with lower energies. For example, imagine that in Figure 7.3 the level at 12.5 eV somehow had more electrons than the one at 12 eV. What would be the result?

Einstein showed that if a photon of the same energy as that which is to be emitted hits an atom with the "inverted" population, two photons are emitted. They travel in exactly the same direction and with exactly the same wave phase. The term for this effect is coherence. In the example mentioned above, this two-photon emission would occur if a photon of energy exactly 12.5 eV struck an atom that had one at an energy of 12.5 eV, and which was in an "inverted" energy population.

This process, called stimulated emission, in contrast to the spontaneous emission of ordinary light, does not stop with the two photons. Each of them can in turn stimulate another atom, releasing four photons. There is eventually a chain reaction of light, with a very intense beam of coherent photons being built up in microseconds.

The population inversion, the key to the laser process, is achieved by "pumping" in the energy to the atoms by the flashtube, shown in Figure 7.1. If the wrong wavelengths are excited, or if inadequate energy is pumped, only ordinary incoherent light is obtained.

Because lasers have been in existence a shorter time than the traditional fields of physics, there is less scientific literature on the subject. However, the potential for interesting and useful environmental results is great.

MEASUREMENT

Particles in Dirty Water

Water pollution may be in the form of dissolved substances or in the form of particles. How can the proportions be identified?

One way is to use the experimental setup of Figure 7.4. A laser beam shines into turbid or polluted water, and part of the reflection is sent back by a mirror to a light-detecting system. The reflected light intensity varies with the composition of the pollutants in the water.

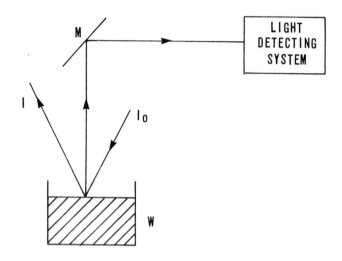

Figure 7.4 Experimental arrangement for detecting particles and water-soluble substances in water. W is turbid water, M is a mirror, I_0 is the incident light beam and I is the reflected light beam.[2]

Why lasers instead of ordinary light? Because it lacks intensity, ordinary light would reflect little from dark water. The coherence and directionality of lasers do not play a part in this experiment.

The problem of measuring particles in contrast to soluble substances is approximated by throwing in bits of Teflon®* to represent the particles, and adding a dark dye (nigrosine) to represent soluble chemicals. The light scatters from the micron-sized pieces of Teflon, and is absorbed by the dye.

Typical experimental results are shown in Figure 7.5. For a fixed density D_T of Teflon particles, the percentage of reflected light increases as the ratio of dye to Teflon decreases. This is what is expected physically—the less dye, the more reflection. However, for a fixed proportion of dye to Teflon density, the ratio of reflected to incident light changes

---

*Registered trademark of E. I. du Pont de Nemours and Company, Inc., Wilmington, Delaware.

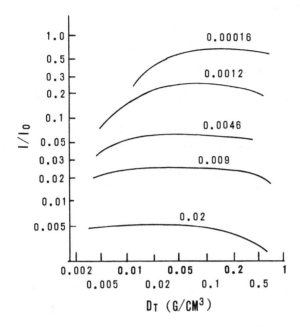

Figure 7.5 Change of the ratio of reflected to incident light as the density of Teflon particles D_T varies. The ratio is plotted as a function of the fraction D_N/D_T appended to each curve, where D_N is the density of dye (nigrosine) in the water. As the density of dye decreases, the intensity of reflected light becomes greater.[2]

only slightly with increased Teflon density. This is shown by the almost flat lines.

If there are many particles in the water and comparatively little dye, D_N/D_T is small. Then I/I_o, the ratio of reflected to incident light, is also small.

By measuring the laser light reflected off water containing different amounts of particles and dye, knowledge is gained of how light reflected off water in rivers and lakes will behave. The technique can be taken out of the lab to find what is in these waters.

Oil and Lasers

Suppose all the Teflon particles in the previous section had floated on instead of in the water. This is analogous to an oil spill, which stays on the surface. Can we use a laser to detect its presence and properties?

164 PHYSICS OF THE ENVIRONMENT

The absorption of laser light in water is described by the Beer-Lambert law which was discussed in Chapter 6:

$$\frac{I}{I_o} = \exp(-za) \tag{3}$$

where z is the thickness of the water or oil, and a is the absorption coefficient, a measure of how fast the light is absorbed with increasing depth. Oil is a much better absorber of light than water. With oil's much greater value of a, the presence of the water underneath can be ignored in the calculations. The experimental setup is similar to that of Figure 7.4. Laser light shines on an oil-covered surface, and the reflected light is measured. The properties of lasers are intensity and directionality. The oil film thickness is measured independently by noting the volume of oil added to the water, and dividing by its area. Results are shown in Figure 7.6.

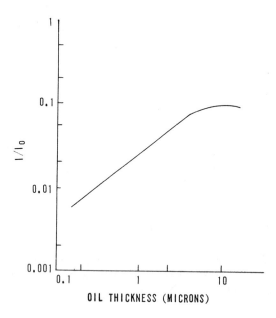

Figure 7.6 Variation of relative amount of laser light reflected from oil on water as a function of oil thickness.[3]

The value of I/I_o rises linearly on the logarithmic scales with increasing oil film thickness. If laser light was directed from an aircraft onto an oil slick, and the percentage of light returning and Figure 7.6

were used, the depth of the oil thickness could be mapped. For example, if 1% of the light were reflected it would indicate an oil film about 3 μ thick. The graph indicates that for a wide range of oil thickness, this method would be useful and accurate.

Air Pollutants

To detect air pollutants, the same principle of absorption contained in Equation 3, the Beer-Lambert law, can be used. Pollutants in air absorb much less light than those in water. A much longer distance, z, is needed to produce the same relative change in I/I_o, the ratio of reflected to original light intensity.

To measure air pollution with lasers, two of its properties are used—intensity and monochromaticity. The latter refers to its extremely narrow range of frequencies, bandwidth, as compared to ordinary light. For example, a laser can have a bandwidth as small as 10^{-7} Å whereas a narrow line in an ordinary spectrum has one of the order of 10^{-2} Å, a factor of 100,000 wider.

Each air pollutant has a definite absorption coefficient. But sometimes pollutants have similar coefficients. To tell them apart, a laser's frequency can be changed over a short range by certain experimental techniques. Its frequency can then be made one which the pollutant absorbs strongly, producing differentiation between pollutants. In Table 7.1, for example, the carbon dioxide laser used to measure absorption for ethylene, ozone and other gases can be changed in wavelength from 9.48 to 10.76 μ. This is a carbon dioxide laser instead of ruby.

The absorption coefficient has been written as a' instead of a. Coefficient a has been expanded into two factors:

$$a = a'c \qquad (4)$$

where a' is the absorption coefficient per unit concentration of absorber (or pollutant) in cm^{-1} atm^{-1}, and c is the actual concentration in atmospheres (atm). Note that the units of $a'c$ are still those of a. When c is known, the concentration of the pollutant is known. In air pollution experiments, the pollutants often have concentrations of a few parts per million of atmosphere.

To find c, Equation 3 is written with the modification of Equation 4 as

$$\frac{I}{I_o} = \exp -za'c \qquad (5)$$

c is then found. In this way, concentrations of common air pollutants can easily be determined down to at least the parts per million range.

Table 7.1. Absorption Coefficients for Different Laser Wavelengths[4]

Pollutant	Laser Type	Wavelength (μ)	Absorption Coefficient per Unit Concentration of Gas at 1 atm, a' (cm^{-1} atm^{-1})
Ammonia	CO_2	10.23	15.4
		10.72	32.0
		10.74	15.8
		10.76	4.8
Ozone	CO_2	9.48	10.8
Sulfur Dioxide	Neon	7.43	15.9
Nitric Oxide	Iodine	5.50	1.2
Ethylene	CO_2	10.23	5.0
		10.53	36.0
		10.65	4.5
Methanol	CO_2	9.66	17.0

CONTROL

A laser technique which will undoubtedly be used extensively in future environmental studies is holography.

Holograms of the Air

Determining size distribution is one problem in identifying particulate matter in the air. The health effects of small particles often differ from those of large ones. One standard way to find this distribution is to let the dirt fall on a white sheet and then measure the sizes under a microscope. This indicates the distribution as the particles hit the ground, but nothing about the distribution in the air, near the smokestack or other sources of pollution. It would be useful to have a three-dimensional "snapshot" of the particles as they came out of the stack, so that their size distribution could be studied at leisure. The hologram can do this.

The ideas behind holography are shown in Figure 7.7. In part A, the laser beam falls on the object to be holographed at the point of the triangle. An image of the object falls on the photographic plate. The plate is represented by the vertical line. By itself, this process would not produce even a two-dimensional image when developed. To produce the hologram, part of the original laser beam is shone on the plate, so that this "pure" beam arrives at the same time as the "image" beam.

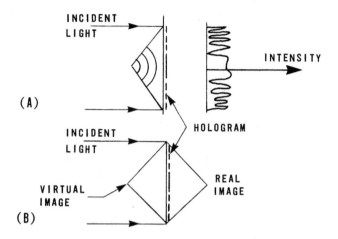

Figure 7.7 Schematic diagram of the holographic recording and reconstruction of a pointlike scattering center. The intensities shown are of the hologram.[5]

When two coherent light beams of the same wavelength overlap, as they do when they reach the plate, they interfere. If the two beams are completely in phase, they produce a maximum, or a bright spot. If they are completely out of phase, they cancel each other and produce a dark spot. The object being holographed puts part of the laser beam which strikes it into different phases with respect to the original beam.

When the plate is developed, a pattern of dark and light images is produced, similar to the water pattern when pebbles are dropped into a pool. If a hologram is taken of an object with recognizable features, like a chess piece, all that is seen is a series of concentric circles and rings of various diameters.

The way to obtain the three-dimensional reconstruction of the original image is shown in part B. Another laser is beamed through the plate from the same direction as the original "pure" beam. Two images are obtained: a virtual one, behind the plate, and a real one, which can be viewed in three dimensions. The holographic image is produced by the laser beam and the previous "inside-out" image.

Holography can be applied to air particles. Although not portrayed here, the hologram would show concentric dark and light circles, each group representing a particle.

168 PHYSICS OF THE ENVIRONMENT

The radius of the central dark circle of each set is approximately proportional to the radius of the particle it represents. This gives an idea of size distribution. Figure 7.8 shows a typical distribution of air particles measured this way. There is a maximum around 3 μ, and a rapid dropping off with increasing size.

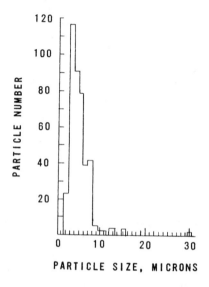

Figure 7.8 Holographically determined size distribution of air particles.[5]

The hologram captures an environmental problem so it can be seen, literally, from different angles. It can be used to determine air particle sizes. With this information, the source and possible harmful effects to humans can often be identified.

Measuring Air Pollutants Directly

Many methods of measuring pollution by lasers are indirect. This approach is often chosen because the direct method is usually more difficult. An example of this is shown in Figure 7.9, which depicts the absorption of laser light in air containing the pollutant nitrous oxide (NO), for a short range of frequencies. Consider trace A, the calibration curve for a measured amount of nitrous oxide. Peaks 2, 3, 4, 7, 9 and 10 are absorption peaks for water vapor, so they are ignored. Only peaks 1, 5, 6, 8 and 11 are due to nitrous oxide. The arrows in trace A indicate what height corresponds to 5 ppm NO.

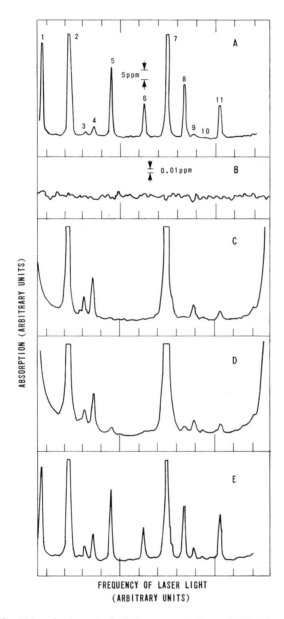

Figure 7.9 Using the laser to find the concentration of NO (nitrous oxide) in air. Trace A is the calibration curve; trace B is a typical curve of noise, or random errors, in the system with the laser turned off; trace C is laboratory air; trace D is air near a highway; and trace E is automobile exhaust. Some peaks are "squared off" because they have reached the maximum value for the particular recording instrument employed.[6]

The amount of detail in Figure 7.9 is great. Laser light absorption is measured for a number of experimental conditions including car exhausts and air near highways. Trace B shows possible random errors. The errors are of the order of 0.03 ppm, much less than expected concentrations. As expected, car exhausts have the highest concentration of NO. This absorption method can be used to detect both individual pollutants and the state of the air as a whole.

SUMMARY

Lasers can be used to learn more about water and air pollution. Its unique properties can be applied to many problems in our environment.

PROBLEMS

7.1 In Figure 7.2, consider a photon being created at point A, near the axis of the ruby. Assume that it bounces once off the left-hand mirror and then exits through the right-hand partially silvered mirror. What is the range of angles it can make with the axis if it is to stay within the beam width of 5 ppm? Find the range of angles for a photon being created at B, at the outer edge of the ruby. What would be the allowable range of angles for photons at A and B if the number of bounces were 10? Note: Approximations to tan θ as θ becomes very small may be required.

7.2 Suppose there were two energy levels E_1 and E_2 to consider, with energies 0 and $4E$, respectively. E is a constant. Plot the relative number of electrons N_2/N_1 as a function of the quantity kT/E as the latter ranges from 0.2 to 20. Describe physically what is happening to the system.

7.3 In Figure 7.5, disregard the nonlinear portions of the curves. Thus for each value of D_N/D_T, I/I_0 is constant. Find one equation which will relate the quantities D_N, D_T and I/I_0 of the four lowest lines. Note that logarithmic scales are used in Figure 7.5. The answer will indicate the range of particle and soluble substance density which a given laser beam can analyze.

7.4 The linearity of the left-hand side of Figure 7.6 is due to the type of scales chosen. Assuming that there is true linearity up to about 5 μ, what is the equation of I/I_0 vs oil thickness? Plot this equation on regular graph paper. Could Figure 7.6 be used for films thicker than about 5 μ? Why does the leveling off occur?

7.5 Suppose a detecting instrument can determine a difference between I and I_0 as low as 1%. If a' is 20 $cm^{-1} atm^{-1}$, how long a pathlength will be required to detect a pollutant concentration of 10 ppm?

7.6 What laser wavelength would be used to find ammonia in the air? Why? How much better, in terms of path-length required to find a given concentration, is it than the worst of the other three wavelengths? Use Table 7.1.

7.7 The size distribution of air particles generally follows what is called a "Poisson distribution." What is this and how closely does Figure 7.9 follow it?

7.8 Another way of considering the variation of air particle size is by means of a "cumulative distribution." It could be said, for example, that 5% of the particles had a diameter less than 2 μ, 12% less than 3 and so on, and these results could be plotted. Do this for Figure 7.8 and find the diameter below which half the particles fall.

7.9 For traces C, D and E in Figure 7.9 measure the height of the 4 NO peaks (excluding peak 1) above background, if possible. Compare them to the height of the corresponding peak of trace A. If trace A represents an NO concentration of 20 ppm, what is the average concentration for traces C, D and E?

REFERENCES

1. Brown, R. *Lasers, Tools of Modern Technology*, (London: Aldus Books, Ltd., 1968), pp. 10, 16.
2. Granatstein, V.L., M. Rhinewine, A.M. Levine, D.L. Feinstein, M.J. Mazurowski and K.R. Piech. "Multiple Scattering of Laser Light from a Turbid Medium," *Appl. Op.* 11(5):1219 (1972).
3. Measures, R.M., and M. Bristow. "The Development of a Laser Fluorosensor for Remote Environmental Probing", presented at Joint Conference on Sensing of Environmental Pollutants, Palo Alto, California, November 8, 1971, paper 71-1121, p. 5.
4. Hanst, P.L. "Spectroscopic Methods for Air Pollution Measurements," *Adv. Environ. Sci. Technol.* 2:136 (1971).
5. Seger, G., and F. Sinsel. "Investigation of an Atomizing Device by Means of Short-Pulse Microholography," *Staub* 30(11):36,38 (1970).
6. Hinkley, E.D. "Tunable Infra-Red Lasers and Their Application to Air Pollution Measurements," *Optoelect.* 4(2):79 (1972).

CHAPTER 8

ONE ATOM, INDIVISIBLE:
ATOMIC PHYSICS AND RADIOACTIVITY

". . . Matter in solid, massy, hard impenetrable, moveable Particles . . . being Solids, are incomparably harder than any porous Bodies compounded of them; even so very hard, as never to wear or break in pieces."

Isaac Newton, *"Opticks"* (17th Century)

"The word atom means that which IS not divided, as easily as it may mean that which CANNOT be divided."

A physicist of 1876

"The lighting effects beggared description. The whole country was lighted by a searing light with the intensity many times that of midday sun. It was golden, purple, violet, gray and blue. It lighted every peak, crevasse and ridge of the nearby mountain range with a clarity and beauty that cannot be described but must be seen to be imagined."

Observer at Alamogordo, New Mexico, 1945

INTRODUCTION

For at least two centuries, the firm belief was held that the atom was truly indivisible. Within last the century, it has become clear that the atom itself is made up of finer particles. The existence of these particles produced, among other things, the quotation of wonder on the part of an observer at the first atomic bomb explosion in 1945. While knowledge of atomic physics and radioactivity has, to some extent, opened a Pandora's box of problems, understanding the principles of the field can shut it, at least partially, again.

One of the greatest controversies in the last few years has been raised by the most prominent peacetime example of atomic physics—nuclear

174 PHYSICS OF THE ENVIRONMENT

power reactors. Lack of space prevents evaluation of the many ways these reactors might affect the environment. Considering their radiation shows how environmental questions may be evaluted by the "order-of-magnitude" technique.

Radiation and Reactors

Table 8.1 shows the estimated impact of the four major sources of radiation on the U.S. population in the years 1970 and 2000. The "rem" is a unit of the effect of radiation on the average human, so that a "man-rem" is the equivalent of one person exposed to one rem for a year. The rem is a function of both the type and amount of radiation affecting all human biological systems. The numbers in the table are only estimates. The present estimated exposure from nuclear reactors is far lower than that from the natural background, which comes from uranium and thorium in rocks and cosmic rays from outer space, among other sources. Of the four categories, exposure from reactors is expected to rise the fastest from 1970 to 2000. Since the natural background radiation per person remains constant, the increase in this category shown in Table 8.1 is due to the expected population expansion. It is assumed there will be no catastrophes in nuclear reactors.

Table 8.1 Estimated Impact of Radiation in the United States, 1970 and 2000[1]

Source	Estimated Exposure in 1970 (man-rems)	Estimated Exposure in 2000 (man-rems)
Natural Background	27,000,000	40,000,000
Medical X-Rays	18,000,000	40,000,000
Fallout from Nuclear Weapons Tests	1,000,000	1,600,000
Nuclear Power Reactors	400	56,000

If the exposure from nuclear reactors was to rise continually at the rate shown in Table 8.1, a serious environmental hazard would be produced. But since 30 years is a long period to extrapolate into the future, further extrapolations would be speculation.

Relative exposure from nuclear reactors compared to other types of radiation is shown in Figure 8.1, where the exposures of the four categories of Table 8.1, along with two other groups, are plotted as a function of time. The occupational group includes on-the-job radiation; "miscellaneous" includes, for example, radiation from television, consumer products and air travel.

ATOMIC PHYSICS AND RADIOACTIVITY 175

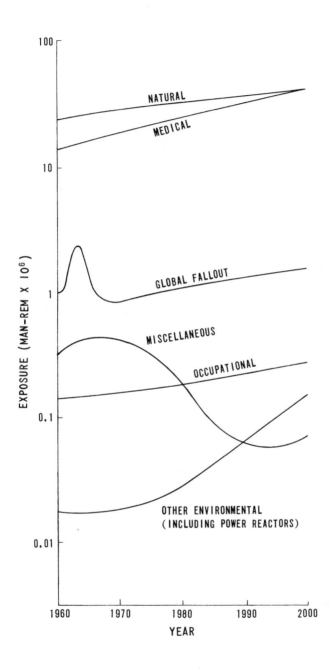

Figure 8.1 Estimated radiation doses in the United States, 1960-2000.[1]

176 PHYSICS OF THE ENVIRONMENT

As in most projections of this type, it is assumed that present conditions will only gradually change. This accounts for the constant slope of most of the lines in Figure 8.1. The slope is due to the projected increase in population.

Radiation from nuclear reactors probably will not reach levels comparable to those from other sources by the turn of the century, although it will rise rapidly for a long time. One major approximation made in this discussion is assuming an average distribution of radioactivity effects. For example, people near a reactor will be exposed to more radiation than those far away. These effects have been averaged out in Table 8.1 and Figure 8.1. Whether any radiation should be allowed from reactors or not is a political, economic and even moral question, but at least this discussion gives a better idea of the relative magnitudes involved.

Fallout, North and South

The most obvious peacetime environmental effect of nuclear weapons is the radioactive fallout produced by nuclear test explosions.

Consider how the radioactive debris is distributed between the atmospheres of the northern and southern hemispheres, and how the quantities change as the fission products fall to the earth. A simple model is shown in Figure 8.2. The size of sky-borne fission products in the northern and southern hemispheres can be represented by x and y, respectively. Suppose a fraction K of each is transferred from one to the other by winds in a short time Δt. The units of K are time^{-1}. The amounts transferred north-south and south-north are then Kx and Ky. A fraction k of each falls on the earth, so that the amounts which fall are kx and ky. Finally, a fraction r of x and y disintegrates radioactively in each hemisphere, so that the amounts which disintegrate are rx and ry. k, r and K are found by determining how x and y vary with time.

The quantity gained by the northern hemisphere from the south is Ky. The quantity lost is x(K + r + k), where xK is lost by transfer to the southern hemisphere. The change in x is then Ky - x(K + r + k), the gain minus the loss. This change in x, or Δx, occurs in a time Δt, so

$$\frac{\Delta x}{\Delta t} = Ky - x(K + r + k) \tag{1}$$

where "Δ" indicates a change. The units of both sides are (amount of radioactivity)/time. To find x as a function of time, allow the Δ's, or changes, to become smaller and smaller. In the language of calculus, they become differentials. Then

$$\frac{dx}{dt} = Ky - x(K + r + k) \tag{2}$$

ATOMIC PHYSICS AND RADIOACTIVITY 177

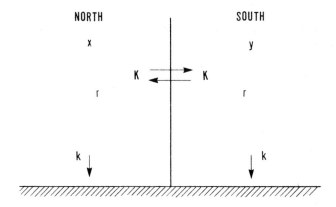

Figure 8.2 Model of transfer and change of radioactivity in the atmosphere, with two compartments representing northern and southern hemispheres. A fraction K of the radioactivity in each hemisphere is transferred to the other by winds, a fraction k falls to earth, and a fraction r disintegrates.[2]

Similarly, for the southern hemisphere,

$$\frac{dy}{dt} = Kx - y(K + r + k) \tag{3}$$

This last equation can be found by interchanging x for y. Solving these two equations takes advanced mathematics. It is then found that

$$x - y = (x_0 - y_0) \exp\{-(k + r + 2K)t\} \tag{4}$$

$$x + y = (x_0 + y_0) \exp\{-(k + r)t\} \tag{5}$$

These are the difference $(x - y)$ and sum $(x + y)$ of the amounts in the two hemispheres. The quantities x_0 and y_0 are the values of x and y at time $t = 0$.

K can be found by dividing Equation 4 by Equation 5, getting

$$\frac{x - y}{x + y} = \frac{(x_0/y_0) - 1}{(x_0/y_0) + 1} \exp(-2Kt) \tag{6}$$

where t is in years.

To compare theory with experiment, a particular radioactive isotope is studied. Each isotope has a different rate of disintegration r. One of the most important isotopes is strontium-90, because of its cumulative effect in human bones. Its atmospheric concentration has been sampled for years by means of aircraft and unmanned balloons.

Equation 6, for the isotope strontium-90, is plotted as the lower set of points in Figure 8.3. October 1963, has been taken as the initial time $t = 0$. The graph shows a continual decrease in the level of strontium-90.

178 PHYSICS OF THE ENVIRONMENT

By plotting the points on semi-logarithmic scales, the fraction K of strontium-90 transferred between hemispheres can be found.

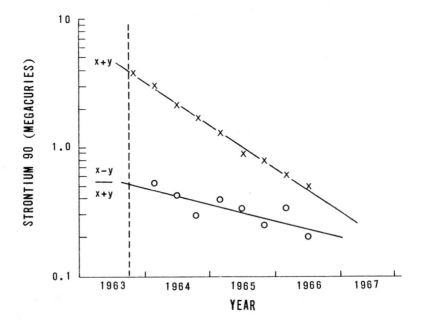

Figure 8.3 Strontium-90 in the atmosphere from 1963 to 1966. x represents the amount in the northern hemisphere; y, the amount in the southern hemisphere. The initial time is taken as October 1963, in the computations.[2]

In the same way, (x + y) of Equation 5 can be plotted to find the sum of (k + r).

By measuring radioactivity in the atmosphere, it can be determined how much is falling, and how much is being transferred between the two hemispheres. This information is used to determine the likely effects of nuclear tests on different regions of the world.

Neutron Activation

How can the physics of the atom be used to measure very small amounts—traces—of potentially dangerous elements in the environment? Measuring mercury levels in fish requires apparatus that can determine concentrations in the parts per million (ppm) range. Methods are available to find elements in the parts per billion (ppb) range. An important technique is neutron activation analysis.

If negative electrons were propelled at a target, they would tend to be repelled electrically by the electron cloud around the nucleus of each atom. A stream of electrons would return. If positive protons were propelled at the target, they would get past the electron cloud, being electrically attracted, but would be repelled by the positive nucleus. The stream of protons would be scattered back. However, if electrically neutral neutrons were propelled at the target, they would get by the electron cloud, being unaffected, and strike the nuclei of the atoms in the target, again being unrepelled and unattracted. Since neutrons are comparatively heavy nuclear particles, having about the same mass as the proton, they often disintegrate the nuclei they strike.

Upon disintegration, a wide variety of particles are given off. The most common are beta particles (electrons) and gamma rays (electromagnetic radiation).

Each element in the target gives off a gamma ray (or beta particle) of characteristic energy when struck by neutrons. For example, element x may have gamma rays of exactly 4.017 units of energy. To determine what is in the target, the energies are measured and checked in reference books.

Suppose that there are N atoms in the target, of which a small number, N^*, have been made radioactive by bombardment with neutrons. How does N^* vary with time?

The radioactivity depends on the number of atoms N present—the more atoms, the more radioactivity. Similarly, the radioactivity will increase linearly with the "cross section" σ of the N atoms, a measure of the ability of the incoming neutrons to make a given type of atom radioactive. The increase in radioactivity is then proportional to σN. The increase in radioactivity after neutron bombardment is then $k_1 \sigma N$, where k_1 is another constant.

Along with the increase goes a decrease, since the radioactive atoms are themselves disintegrating. The number disintegrating in a short period of time is proportional to N^*, the number of radioactive atoms, and to r, the decay constant, a measure of how fast these atoms disintegrate after formation. The decrease in radioactivity is then rN^* because the proportionality constant in this case equals 1. Summarizing,

(change of radioactivity in a short period of time)
= increase of radioactivity-decrease of radioactivity
= $k_1 \sigma N - rN^*$ (7)

The small change in radioactivity is ΔN^*, and the small change in time is Δt. Then

$$\frac{\Delta N^*}{\Delta t} = k_1 \sigma N - rN^* \qquad (8)$$

Allowing the time and change of radioactivity to become small,

$$\frac{dN^*}{dt} = k_1 \sigma N - rN^* \qquad (9)$$

The term $k_1 \sigma N$ is affected only slightly by changes in N^* and t. The number of N (nonradioactive) atoms is always much larger than the number of N^* (radioactive) atoms. For all practical purposes, then, N is a constant. Since k_1 and σ are also constants, the three may be multiplied together to yield the constant k_2. Then

$$\frac{dN^*}{dt} = k_2 - rN^* \qquad (10)$$

This equation is roughly similar to Equations 2 and 3. When integrated,

$$k_2 - rN^* = k_3 \exp(-rt) \qquad (11)$$

where k_3 is yet another constant.

When the neutron bombardment of the target is just beginning, or when time t is zero, the number of radioactive atoms N^* is zero. Then Equation 11 becomes

$$k_2 = k_3 \qquad (12)$$

Substituting this back into Equation 11,

$$rN^* = k_2 \{1 - \exp(-rt)\} \qquad (13)$$

This is an expression for the number of radioactive atoms in the sample as a function of time. Its curve can be described as s-shaped or sigmoid.

Results of using neutron activation analysis to detect trace elements are shown in Figure 8.4. The bottom part shows the energy each gamma ray produced from the sample in thousands of electron volts, or kilo-electron volts (keV).

The left-hand side of Figure 8.4 indicates the number of "counts" or disintegrations for a particular energy. Because of the large variation in counts, the scale is logarithmic. Each spike indicates the energy for a gamma ray emitted from a particular element. For example, at about 890 keV there is a spike indicating a gamma ray from scandium-46. By comparing the height of this spike with that produced when the same experiment is performed with a known mass of scandium, the mass of this element in the sample can be found. Gamma ray energies of all elements have been measured and tabulated. The approximate limit of detection of trace elements by neutron activation analysis is shown in Table 8.2 In principle, most elements can be detected at the parts per billion

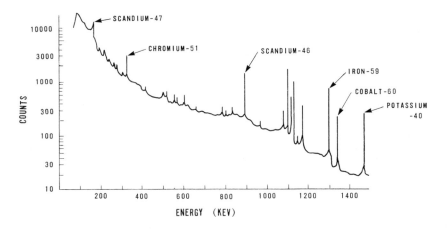

Figure 8.4 Typical gamma-ray spectrum of energy radiated from a sample after neutron activation analysis. Only the major peaks are labelled.[3]

range. This method is not used for all environmental identification problems because molecules rather than elements are often under consideration, and then the gamma-ray spectrum produced becomes even more complicated than Figure 8.4. Secondly, there are comparatively few sources of neutrons strong enough to produce the results of Table 8.2. Thirdly, it may take hours of bombardment to produce enough counts to tell the peaks from the background. However, under the right conditions, neutron activation analysis is one of the most sensitive methods for determining pollution levels.

Water and Deuterium

New sources of freshwater are becoming fewer and fewer. With the increasing pressure to preserve water resouces has come the concept of "closed-loop" water supply systems. In these systems, used water would be recycled into clean water. The idea has many advantages from the conservation point of view, but there is one less promising aspect worth discussing: will the proportion of deuterated water in such a system increase rapidly?

The natural isotope of hydrogen, deuterium, has an atomic weight of two instead of the usual one. It can form what is called deuterated water, with the chemical formula HDO, where D represents an atom of deuterium which substitutes for a regular hydrogen atom. In a 5-gal bucket of ordinary water, about 0.2 oz (300 ppm) will be deuterated water. A

Table 8.2. Limits of Detection for 75 Elements by Neutron Activation Analysis[3]

Detection Limit (ppb)	Elements
0.00003	Dy
0.00009	Eu
0.0003	—
0.0009	Mn, In, Ln
0.003	Co, Rh, Ir
0.009	Br, Sm, Ho, Re, Au
0.03	Ar, V, Cu, Ga, As, Pd, Ag, I, Pr, W
0.09	Na, Ge, Sr, Nb, Sb, Cs, La, Er, Yb, U
0.3	Al, Cl, K, Se, Kr, Y, Rn, Gd, Hg, Tm
0.9	Si, Ni, Rb, Cd, Te, Ba, Tb, Hf, Ta, Os, Pt, Th
3	P, Ti, Zn, Mo, Sn, Xe, Ce, Nd
9	Mg, Ca, Tl, Bi
≐	F, Cr, Zr
90	Ne
300	S, Pb
900	Fe

substantial increase in this proportion can lead to health problems. Proliferation of closed-loop water recycling systems would then be avoided.

H_2O has a higher vapor pressure, or evaporates faster, than HDO. The more evaporation, the greater the proportion of HDO to H_2O. If this occurs over centuries, there is nothing to worry about. If it takes weeks or months, there is cause for concern.

Let the system have a capacity or volume S. Its rate of water loss by evaporation is M, and this is replaced by new outside water of the same volume. Let the concentration of HDO in this outside water be D_o, and the concentration of HDO in the system's water at any time t be D_t.

The amount of HDO added to the system from the outside water is the volume of this water times its HDO concentration, or MD_o. Meanwhile, HDO is being lost from the system through evaporation. This loss equals the volume of water loss (M) times the actual concentration of HDO in the system (D_t) times the relative vapor pressure of HDO to H_2O, denoted by x. This last factor gives a measure of how fast the HDO is evaporating. The rate of HDO loss by evaporation is then $MD_t x$, where x is less than one. The net rate of increase of HDO is given by the addition minus the loss, or $M(D_o - D_t x)$.

Determining the change in concentration implies division by the capacity S of the system, obtaining $(M/S)(D_o - D_t x)$.

Using the same reasoning as that leading up to Equation 1 and 8, the change of HDO concentration in a short time Δt is

$$\frac{\Delta D_t}{\Delta t} = \frac{M}{S}(D_o - D_t x) \qquad (14)$$

In this equation, D_t is the independent variable, or the analog to y. The Δ's become differentials, producing

$$\frac{dD_t}{dt} = \frac{M}{S}(D_o - D_t x) \qquad (15)$$

This equation is similar to Equation 10, which was solved before. All the other terms except t and D_t are constants.

Thus, it is found that

$$D_t = \frac{1}{x}\{D_o - K \exp(-xMt/S)\} \qquad (16)$$

where K is a constant. When plotted, this is an s-shaped or sigmoid curve. At the start of the experiment, the concentration of HDO inside the system is the same as outside. This can be written as $t = 0$, $D_t = D_o$. Substituting this condition into Equation 16 to find K,

$$D_o = \frac{1}{x}(D_o - K), \text{ since } e^o = 1$$

Then

$$D_o x = D_o - K$$

and

$$K = D_o(1 - x) \qquad (17)$$

Substituting Equation 17 into Equation 16, the variation of D_t with time t is then

$$D_t = \frac{D_o}{x}\{1 - (1 - x)\exp(-xMt/S)\} \qquad (18)$$

This equation shows how the deuterated water concentration changes with time.

Figure 8.5 shows an analogous study done on tritiated water (HTO) as dry air flowed across the water surface. T is the symbol for tritium, or the isotope of hydrogen with an atomic weight of three. The solid line is a calculated curve. The concentration of tritiated water rises approximately 20% over a nine-day period. This poses a strong health hazard because of

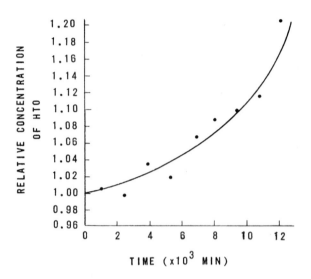

Figure 8.5 Change of the concentration of tritiated water with time. The initial concentration is given the value 1.00.[4]

the dangerous radiation that tritium gives off. The magnitudes involved in a closed-loop system can now be calculated.

MEASUREMENT

Half-Lives

Radioisotopes disintegrate or decay at differing rates—depending on their half-lives. If a particular isotope has a half-life of a microsecond, it may be difficult to record its existence; if its half-life is in years, no problem is posed. (See Figure 8.6). Trace elements in air pollutants were being measured by neutron activation analysis. The shape of the gamma-ray curve varies dramatically with time. Some peaks which were barely apparent in the first curve, taken 3.9 min after the neutron bombardment ended, are prominent in the last, taken 320 min after irradiation.

As well as appearing with time, peaks can also disappear. For example, there is a peak at 1.43 million electron volts (MeV) which virtually vanishes by curve C, or 33 min after irradiation. The isotope which produced the peak, vanadium-52, has a half-life of 3.8 min.

The intensity of every radioactive source eventually decreases to a low level. Sometimes the wait lasts a few thousandths of a second. Other times, it will be much longer than a human lifetime.

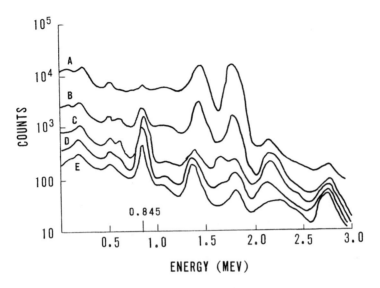

Figure 8.6 Change in the gamma-ray spectrum of air pollutants with time. Curves A, B, C, D and E denotes times 3.9, 11.7, 33, 120 and 320 minutes, respectively, after irradiation.[5]

Sewage and Radioactivity

Sewage dumping into bays and oceans has become an environmental problem in the past few decades. Using radioactive tracers is useful in finding how wastes are distributed. The waste is "tagged" with radioactivity, then the water is tested for the tracer in the vicinity of the sewer outflow. A typical example of how the affected area changes with time is shown in Figure 8.7. Only the extent of the area where radioactivity was detected is shown. The tracer used was scandium-46. The area covered by waste for this particular system increases rapidly at first and then slows down.

Figure 8.7 gives a reasonable indication of how fast sewage could spread. For example, about 60 km² are covered in a year. Sewage can diffuse over a wider area than first expected.

CONTROL

Elimination of Bacteria

Doses of radioactivity can be used to sterilize sewage. One measure of sterilization deals with the biochemical oxygen demand (BOD). Large amounts of organic wastes use up large amounts of dissolved oxygen, so

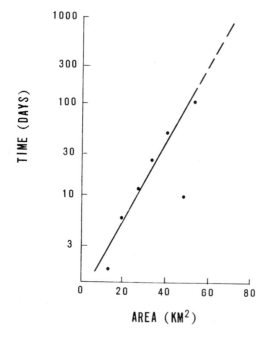

Figure 8.7 Spread of radioactive tracer as a function of time on a seabed.[6]

usually the greater the degree of pollution, the greater the BOD. In the experiment, high-BOD sewage was irradiated. The process not only lowered the BOD, but lightened the sewage color. There are then two measures of the effectiveness of the process: the BOD and the color. The latter measure can be put on a more quantitative basis by finding the change in transmission of light passing through the sewage water.

Typical results are shown in Table 8.3. As the irradiation increases, the BOD of the sewage decreases, indicating the effectiveness of the treatment. Since measuring the BOD is a lengthy and expensive process, the simple measurement of light transmission can be used if it correlates well with BOD. In the table, the rad is a unit of radioactivity and a measure of the energy produced by the irradiation per unit mass of the sewage.

Table 8.3 shows that both the BOD and the light transmission fall with increasing irradiation. The method has potential as a way to clean up sewage.

Planning Sewage Systems

Liquid wastes tagged by radioactivity have already been discussed. To simplify matters, it was assumed that the waste spread uniformly from the outlet pipe into the sea. In practice, this often is not true. Currents

Table 8.3 Change of Light Transmission and
Biochemical Oxygen Demand (BOD) of Sewage with Gamma-Ray Irradiation[7]

Sample Number	Irradiation (rads)	BOD (ppm)	Light Transmission at Indicated Wavelength (%)			
			2,800 Å	4,000 Å	5,000 Å	6,250 Å
Control	0	202	0	22	48	58
1	2×10^6	128	0	15	43	52
2	4×10^6	101	0	12	41	51
3	6×10^6	85	0	5	26	41
4	1.2×10^7	76	0	8	36	48
5	2.4×10^7	53	0	5	26	37

will concentrate more waste in one area than another. These changes in concentration must be known when planning sewer systems.

Consider a proposed sewer outlet that will have a discharge rate q. Suppose a radioactive tracer solution of concentration C were also to be discharged from that outlet. If enough time passed for equilibrium under the action of tides and currents, the concentration C_T of the tracer in samples taken offshore would be proportional to both C and q:

$$C_T = k_1 Cq \tag{19}$$

where k_1 is a constant. The quantities C and q are both known.

Instead of a radioactive concentration, consider a pollutant in a sewer discharge of rate Q. Suppose this new concentration of pollutant is C_o. Using the reasoning employed in Equation 19,

$$C_p = k_2 C_o Q \tag{20}$$

where C_p is the pollutant concentration in offshore samples and k_2 is another constant. Equations 19 and 20 both describe the same flow situation, with only the material in the water or sewage changing. Then write $k_1 = k_2$.

Since the projected pollutant concentration is to be compared to the measured tracer concentration, define

$$D = \frac{C_p}{C_T} \tag{21}$$

where D is the dilution ratio. In those areas where D is large, a large pollutant concentration is to be expected; where it is small, the opposite occurs. C and q can be measured and C_o and Q estimated, so that D is

188 PHYSICS OF THE ENVIRONMENT

$$D = \frac{C_o Q}{Cq} \tag{22}$$

Figure 8.8 is a map showing a typical proposed sewer outlet in Norway. Each of the sampling points is indicated by a dot. To simplify the map, the values of D were grouped into three ranges, which can be called high, medium and low anticipated pollution. The highest value of D is near the proposed sewer outlet, but the way in which D varies is not completely foreseen. For example, point A has a lower ratio than expected on the basis of distance from the outlet.

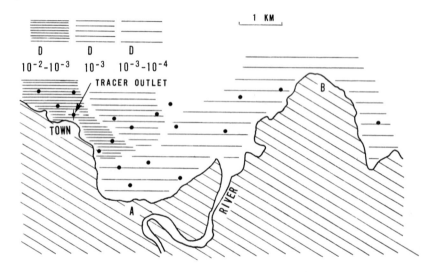

Figure 8.8 Distribution of dilution ratio D near a proposed sewer outlet in Norway. The dots indicate sampling points. As expected, the highest values are near the outlet, but the geography of the area influences the distribution of D.[8]

By calculating the dilution ratio, a reasonable idea of where water pollutants are going to be dispersed can be obtained. If this technique had been used more in the past, many environmental disasters could have been avoided.

Types of Oil

Suppose the Coast Guard comes across an oil slick a few kilometers offshore without a ship in sight. The offender can be found by identifying the type of oil spilled, and then determining which ships in the vicinity carried that variety. The result of one experiment is shown in Table 8.4.

Table 8.4 Ratios of Gamma-Ray Peak Intensities
from Oil Slick Samples in Baltimore Harbor[9]

Peak Ratios	Sample Number			
	1	2	3	4
Tests of 1970				
$\dfrac{492 \text{ KeV Ni}}{1{,}434 \text{ KeV V}}$	0.20 ± 0.10	0.26 ± 0.09	0.38 ± 0.13	0.32 ± 0.11
$\dfrac{1{,}632 \text{ KeV Na}}{1{,}434 \text{ KeV V}}$	3.4 ± 0.4	3.4 ± 0.4	5.7 ± 0.6	3.4 ± 0.4
Tests of 1971				
$\dfrac{492 \text{ KeV Ni}}{1{,}434 \text{ KeV V}}$	0.17 ± 0.03	1.1 ± 0.3	0.80 ± 0.29	0.16 ± 0.04
$\dfrac{1{,}632 \text{ KeV Na}}{1{,}434 \text{ KeV V}}$	2.2 ± 0.1	33 ± 4	22 ± 4	1.8 ± 0.1
$\dfrac{1{,}779 \text{ KeV Al}}{1{,}434 \text{ KeV V}}$	0.035 ± 0.013	0.19 ± 0.07	0.25 ± 0.09	0.054 ± 0.014

Samples of four oils found in Baltimore Harbor were heated till only a residue remained. The ash was then bombarded by protons with energies ranging from 1 MeV to 1.8 MeV. Protons were used instead of neutrons because many elements in the oil residues produce more disintegrations under proton rather than neutron irradiation. Although many particles are given off from the ash, gamma rays are the simplest to measure.

In a gamma-ray spectrum there could be dozens of peaks of interest, so only the ratio of the intensities, or counts, for a few elements have been chosen. The first ratio in Table 8.4 is a comparison of the 492 kiloelectron volt (keV) peak of nickel to the 1,434 keV peak of vanadium. Both elements exist in small concentrations in oil. Other substances considered in this experiment were sodium (Na) and aluminum (Al).

The numbers in Table 8.4 indicate the ratios of the counts for the two peaks considered. For example, the first result

$$\frac{492 \text{ keV Ni}}{1{,}434 \text{ keV V}} = 0.20 \pm 0.10$$

could mean that the 492 keV peak for nickel had about 60 counts, and the 1,434 keV peak for vanadium had about 300. The ± sign in front of 0.10 indicates the standard deviation, or an approximation of the error in

Table 8.5 Ratios of Gamma-Ray Peak Intensities in
Oil Samples from Different Areas[9]

Peak Ratios	Area			
	Huntington Beach, California	Texas	Baxterville, Mississippi	Bay Marchand, Louisiana
$\dfrac{1{,}434 \text{ KeV V}}{492 \text{ KeV Ni}}$	2.1 ± 0.5			
$\dfrac{1{,}779 \text{ KeV Al}}{492 \text{ KeV Ni}}$		0.71 ± 0.29		0.17 ± 0.06
$\dfrac{1{,}632 \text{ KeV Na}}{492 \text{ KeV Ni}}$	33 ± 6	32 ± 8	5.3 ± 0.8	0.20 ± 0.07

the measurement. If the measurement were repeated many times, about two-thirds of the values would fall within one standard deviation of the average. In this case, two-thirds of the measurements fall between 0.10 and 0.30.

Table 8.4 can be used to characterize each sample of oil, and then to compare them to samples taken from different oil fields. Since the origin of the oil on board a ship is usually known, the vessel guilty of dumping or spilling can be found.

SUMMARY

In this chapter, aspects of atomic physics and the environment, ranging from the effects of the bomb to tracking down oil spillers have been covered. Some of the physics has been relatively simple, such as the calculations of the increase of reactor radiation, but other parts have been more complicated, as in the interpretation of gamma-ray energy spectra. This branch of physics can measure and solve many environmental problems.

ATOMIC PHYSICS AND RADIOACTIVITY 191

PROBLEMS

8.1 Suppose that the average rate of increase of exposure from reactors from 1970 to 2000 continues indefinitely. In which year would the exposure reach 500,000 man-rems? 5,000,000 man-rems?

8.2 What accounts for the large variation in Figure 8.1 in the exposure from global fallout between 1960 and 1970? Why is the miscellaneous group expected to decrease for most of the period depicted?

8.3 In order to construct the simple model of interhemispheric radioactivity transfer, many approximation were made. Suggest a few.

8.4 Find the value of K from the slope of the line through the points of Figure 8.3. Its units are in yr^{-1}. Would this value have changed very much if the line had been tilted somewhat?

8.5 Using the top line of Figure 8.3, find k + r, in yr^{-1}. Suppose no new strontium-90 had been added to the atmosphere after 1967. What would the world total be now if the 1963-67 trends had continued?

8.6 Using Figure 8.3, determine the ratio of strontium-90 radioactivity in the northern hemisphere to that of the southern hemisphere for the years indicated.

8.7 The decay constant r is defined by

$$r = \frac{0.693}{T_{1/2}}$$

where $T_{1/2}$ is the "half-life": half the sample of radioactive atoms disintegrate in that time. If the target was bombarded for a very long time, the number of radioactive atoms would be k_2/r, according to Equation 13. What would the number be if it were bombarded for periods of 0.25, 1, 2, 3, 5 and 10 times the half-life of the radioactive atoms? Let $k_2 = 10^6$ sec^{-1} and r = 1 sec^{-1} for this problem.

8.8 At what time does the number of radioactive atoms, as expressed in Equation 13, equal 99% of the ultimate number?

8.9 Some of the spikes in Figure 8.4 do not appear too distinct from the background or smooth curve due to the logarithmic scale. Measure the ratio of the counts for the spikes to those for the adjacent background for some elements such as scandium-46, chromium-51 and potassium-40. What are the advantages and disadvantages of the logarithmic scale?

8.10 A small closed-loop water system has a capacity S of 1,000 cubic meters. The rate of loss by evaporation was 2.5 liter/minute. The relative vapor pressure of HDO with respect to H_2O (x) is about 0.92. If D_0 is 200 ppm, plot D_t as a function of time. How long does it take D_t to get halfway toward its ultimate value?

8.11 Using Figure 8.5, what is the "time constant" Mx/S for the particular experiment shown? The time constant is a measure of how fast the proportion of tritiated water reaches an equilibrium level.

8.12 One of the prominent peaks in Figure 8.6 is at 0.845 MeV (mega- or million electron volts). Subtract the counts for the adjacent background, or continuous curve, from the maximum number of counts at each time. The background can be approximated by taking it as the middle of a line drawn between the two ends of the peak. For example, for curve E the background count for the 0.845 MeV peak would be midway between points F and G, or about 100 counts. Disregard curve A, since a measurement there would be inaccurate. What is the half-life of this particular isotope?

8.13 What proportion of the radioactivity of vanadium-52 in Figure 8.6 is lost between curve A and curve C?

8.14 Find the equation of area covered in Figure 8.7 as a function of time, using the semi-logarithmic nature of the graph.

8.15 What accounts for the fact that the first point of Figure 8.7, at 1.5 days, is not on the straight line?

8.16 For Table 8.3, plot (a) BOD vs dose (or irradiation), (b) percent light transmission vs dose (for the four wavelengths measured), and (c) BOD vs percent light transmission (again for the four wavelengths). On what basis is there a relationship between BOD and irradiation?

8.17 Why is the value of D around point A in Figure 8.8 comparatively low, when it is so close to the outlet? Is it valid to show D as being in the low range around point B, where there are no sampling points nearby?

8.18 Use Table 8.4 to find the four similar samples. Plot each of the four values for each ratio on graph paper. The standard deviations are usually plotted as bars above and below the actual value. For example, in the first ratio the bars would extend from 0.20 + 0.10 (= 0.30) to 0.20 - 0.10 (= 0.10).

8.19 Table 8.5 shows ratios of peak intensities for oil samples from four states. On the basis of the previous problem, which of the 1971 samples 1, 2, 3 or 4 came from Texas? To do this, transform ratios: in other words, (Na/V)/(Ni/V) = Na/Ni. If x = y/z, and S_x, S_y and

Sz are the standard deviations of the respective variables,

$$\frac{S_x}{x} = \sqrt{\left[\frac{S_y}{y}\right]^2 + \left[\frac{S_z}{z}\right]^2}$$

REFERENCES

1. Lieberman, J.A. "Ionizing-Radiation Standards for Population Exposure," *Physics Today* 24(11):36,37 (1971).
2. Peirson, D.H. "Interhemispheric Transfer of Radioactive Pollution from Nuclear Explosions," *Phil. Trans. Royal Soc. London,* 265A (1161): 296,297 (1969).
3. Bhagat, S.K., W.H. Funk, R.H. Filby and K.R. Shah. "Trace Element Analysis of Environmental Samples by Neutron Activation Method," *J. Water Poll. Control Fed.* 43(12):2415,2417 (1971).
4. Horton, J.H., J.C. Corey and R.M. Wallace. "Tritium Loss from Water Exposed to the Atmosphere," *Environ. Sci. Technol.* 5(4):341 (1971).
5. Brar, S.S., D.M. Nelson, E.L. Kanabrocki, C.E. Moore, C.D. Burnham and D.M. Hattori. "Thermal Neutron Activation Analysis of Particulate Matter in Surface of the Chicago Metropolitan Area," *Environ. Sci. Technol.* 4(1):51 (1970).
6. Wright, M.R. "Tracking Waste by Radioactivity," *Hydrospace* 4(5): 42 (1971).
7. Case, F.N., D.L. Kau, D.E. Smiley and A.W. Garrison. "Radiation-Induced High-Pressure Oxidation of Process Effluents," presented at Nuclear Methods in Environmental Pollution, paper IAEA-SM-142a/52, International Atomic Energy Agency, Vienna (1971).
8. Dahl, J.B., U.H. Haagensen, J. Thomassen and O. Tollan. "Water Pollution Studies by Means of Inactive Indium Tracer and Activation Analysis," presented at Nuclear Methods in Environmental Pollution, paper IAEA-SM-142a/43, International Atomic Energy Agency, Vienna (1971).
9. Mandler, J.W., J.H. Reed and R.B. Moler. "Oil Slick Identification using Charged Particle Activation Techniques," presented at topical meeting of the American Nuclear Society on Nuclear Methods in Environmental Research, Columbia, Missouri, August 23, 1971.

ANSWERS TO SELECTED PROBLEMS*

Chapter 1

1.2 Viscosity at 22°C, viscosity at 30°C, index of refraction, specific gravity, surface tension, relative dielectric constant and thermal conductivity.

1.4 Colorimetric method; ions

1.7 5.7×10^{10} kg

1.9 For 1880 and ten-year intervals subsequent, 0, 0.048, 0.088, 0.119, 0.145, 0.163, 0.177, 0.187, 0.195, 0.199 and 0.203

1.13 15.4, 3.35 times the random error

1.14 The random error is 7.4, the total error is 15.3 and the systematic error is 13.4.

Chapter 2

2.4 0.84

2.9 (d)

2.10 $y = 0.0045/x; y = 0.014/x$

2.14 Advise buying the machine since it is better than the one in the text

2.16 About 4.4 cm/sec

Chapter 3

3.1 $W = W_0 10^{(L_w/10)}$

3.2 0.032 w/m², 105.0 dB

3.3 0.00796 w/m², 99.0 dB

3.4 The difference for 100, 200, 600, 800, 1,000, 2,000, 4,000 and 6,000 Hz is about 22.9, 21.9, 25.7, 25.7, 24.8, 22.9, 47.6 and 50.5 dB, respectively.

*Complete lists of answers available from author.

196 PHYSICS OF THE ENVIRONMENT

3.5 108.8 dB

3.6 2×10^{-6} w/m²

3.7 81.5 dB

3.9 Infinite

3.10 The minimum response is -6.7 dB around 3,000 Hz. The difference between the two curves is greatest around 100 Hz 31.6.

3.12 1.04

3.14 About 0.74%

3.15 Average error is 0.011%

3.16 The sensitivity of f with respect to M_* is, in general, $-(1/4\ell)[F/(M + M_*)^3]^{1/2}$

3.18 12.3 mg

3.19 Curve 2 has the greater sensitivity.

3.20 About 1,500 m/sec; f = 150, 300, 450. . . . $\sec^{-1}$

3.21 $7 \times 10^6/m\ell$; $80 \times 10^6/m\ell$

3.25 D = (log M)/6 - constant

3.27 Insulation is 10 log [(x + y)/x]

3.28 About 31 dB

Chapter 4

4.4 It would double.

4.9 $T = \frac{1}{K_2} \ell n \left[\frac{K_1 + E}{K_1} \right]$

4.11 The left side is cooler.

4.16 Add the cream immediately.

4.17 Approximate midmonth temperatures are 20.5°, 25.5°, 25.2°, 23.2° and 11.9°. The greatest change is from mid-September to mid-October.

4.19 Slopes for mid-July, August, September and October are about 0.59, 0.60, 1.03 and 0.34×10^{15} joules per month, respectively.

4.20 Slopes for mid-June, July, August, September and October are about 0.7, 0.4, -1.1, -2.7 and -0.9×10^{15} joules.

4.21 The differences for June 15, July 1, July 15, August 1, August 15, September 1 and September 15 are about 0.16, 0.56, 0.44, 0.48, 0.48, 0.52 and 0.72×10^{15} joules, respectively.

4.23 5Z, if Z is the electrical energy generated

ANSWERS TO SELECTED PROBLEMS 197

Chapter 5

5.1   About 50/min.

5.4   Relationship cannot be linear.

5.5   Up to about 90 ppm

5.6   345, 1,380, 690, 1.2 and 46 ppm P_2O_5; 740, 3,740, 1,750, -1 and -30 μmho/cm

5.8   0.24, 0.32, 0.43, 0.59, 0.71, 0.91, 1.1 and 1.4 x 10^{16} ohm-cm

5.9   For the upper set of points, y = 4.19 - 0.035(t - 1900), where y is the conductivity in 10^{-10} μmho/cm and t is the time in years. The variance of the slope is 7.5 x 10^{-5}.

5.10 The equation of the upper line is y = 3.7 - 0.025(t - 1900) where the quantities are as described in the previous answer. The equation of the lower line is y = 3.1.

5.11 2008

5.12 Using the years of the figure, the numbers are about 1.5, 0.8, 1.7, 2.2, 3.0, 2.4, 3.2, 3.4 and 4.8 x 10^3 nuclei/cm^3.

5.14 From Equation 4, -0.019 and -0.047; from Equation 9, -0.020 and -0.050

5.15 -0.18 μg/Hz

5.16 About 38 and 56 μg; the greatest rate is around 25 hours from the beginning. The rate is around 36 μg/hr.

5.18 About 3,000 nuclei/cm^3; about 3.1 x 10^7 nuclei/cm^3

5.19 About 0.4 volts; about 0.1 msec; about 0.3 msec

5.21 The equation seems to be valid between 40 and 100 microns; both about 0.018 units/micron

5.27 From about 20 ohm-meters to 300; from about 1.5 mg/ℓ to 80; from about 3 ohm-meters to 100; from about 15 mg/ℓ to about 3,000.

5.28 The average resistivity of the top layer in the southern region is about 240 ohm-meters; the next layer down is about 95. For the northern section, the corresponding averages are about 80 and 78. The lowest layer in the northern region is over 500. About 18 mg/ℓ.

5.31 About 130 and 215%. Both have a diameter of about 0.25 km.

5.32 3.2 mm

5.34 0.5 cm/sec; about 11 times

5.35 Maximum is around 180 cm/sec; 332 μamp

5.36 12°C

198 PHYSICS OF THE ENVIRONMENT

5.37 Probably around 2-3 cm/sec
5.39 Around 0.5 cm/sec; approximately 80%
5.43 0.43, 0.69, 0.96 and 1.21 volts; -0.34, 0.40, 0.25 and 0.76 volts
5.45 For a volume of 400 cm^3, line E
5.46 The ratios for 200 and 400 cm^3 are 1.5 and 2.6, respectively
5.49 97.5% and 96.2%

Chapter 6

6.2 The averages are, in percent, 50, 68, 74, 59, 32 and 68
6.3 The positions are 10.4, 10.6 and 10.7 x 10^{4}Å for A, B and C, respectively.
6.4 $100, $105.00, $110.25, $115.76, $121.55, $127.63, $134.01, $140.71, $147.75 and $155.14, assuming an initial value of $100.
6.5 0.58 and 0.75 cm
6.6 Maxima around 8 and 12 x 10^{4}Å
6.7 $(1 - R_s)(1 - R_p) \exp(-KD)$
6.8 0.57, 0.71 m
6.13 $y = 6.5 \exp(2.9x)$, where y is the chlorophyll level in mg/m^3, and x is the signal in volts.
6.14 About 8 and 0.04 mg/m^3; if the signal changes from -1.8 to -1.7 volts, 0.02 mg/m^3
6.16 Pyramid Lake; Lake Berryessa
6.17 40
6.20 Stations 2 and 3; Stations 7 through 12.
6.27 Peak 5
6.29 Six
6.30 None. Only four pairs (1-18, 7-13, 10-16 and 17-18) have as many as five in common.
6.31 8^8, or approximately 16,777,000
6.32 1.25 x 10^{-5}
6.34 Peak 2; Peak 8; the former
6.36 Counting false alarms as both human and instrumental, there were 12.
6.38 Lags are usually about 5 sec.

ANSWERS TO SELECTED PROBLEMS 199

6.41 For (A) the only nonzero point is the sixth, where T is 0.65. For (E), the values of T for points 1 through 10 are 0.3, 0.3, 0.2, 0.15, 0.1, 0.15, 0.1, 0.1, 0.1 and 0.1. Then K_2 is 4.0, 4.0, 5.4, 6.3, 7.7, 6.3, 7.7, 7.7, 7.7 and 7.7 x 10^{-4}.

6.42 For curves A, B, C, D and E, 1.8 and 9.4; 1.4 and 9.4; 0 and 7.1; 4.1 and 9.4; 3.5 and 9.4, all times 10^{-4}. We have assumed the same values for K_3 and x as in the previous problem. Ratios are 0.19, 0.15, 0, 0.44 and 0.37.

6.44 Around distance notch 6 in curves A and B

6.46 For sulfate, between midnight and 4 am; for sulfite, between 9 and 11 am.

6.48 $y = k_1 \exp(k_2 x)$, where k_1 and k_2 are constants; 0.48

6.50 Around 28 sec

6.53 Generally, $y = 100 \exp(K_2 x)$, where K_2 is -0.074, -0.104, -0.117, -0.143, -0.212 and -0.331, from top to bottom; 41 exposures

6.55 The value of K_2

6.56 Methoxychlor, 14 exposures

Chapter 7

7.1 5 x 10^{-7}, 1 x 10^{-6} radians

7.3 $(D_N/D_T)(I/I_0)^{0.7} = 0.005$

7.4 $(I/I_0) = 0.027\ x^{0.8}$, where x is the oil thickness

7.5 230 m

7.6 10.72 microns. 6.5 x 10^{11}

7.8 About 3.7 microns

7.9 2.4, 2.8 and 17.6 ppm

Chapter 8

8.1 2013 and 2027

8.4 0.15, if the year 1963 is taken as t = 0

8.5 0.76, if 1963 is taken as t = 0; 78 curies

8.6 For the beginnings of the years 1963 to 1967, 4.7, 3.0, 2.1, 1.8 and 1.5

8.7 842,000; 500,000; 750,000; 875,000; 968,000; 999,000

8.8 When t = 6.65 $T_{1/2}$

8.9 For the three elements, the ratios are 5.7, 2.3 and 15

8.10 About 210 days

8.11 About 1.2×10^{-5} min^{-1}

8.12 About 160 min

8.13 97.8%

8.14 $t = 0.73 \exp(0.1A)$, where t is time and A is area

8.18 In 1970, samples 1, 3 and 4 are similar; sample 2 is dissimilar. In 1971, samples 1 and 2 are similar.

8.19 Sample 2 in 1971 tests likely came from Texas.

GLOSSARY

Absolute temperature: An absolute scale of temperature has no minus signs. In this scale, -273°C is absolute zero. Zero Celsius is then 273°. The absolute scale of temperature is then the ordinary Celsius scale plus 273°C.

Absolute zero: The lowest possible temperature, -273°C. In spite of this cold, the molecules of matter still have some energy. While this level has not been achieved, it is possible to get within a few thousandths of a degree in the laboratory.

Absorption, heat radiation: Occurs when waves strike an object directly. Since the object is warmed by radiation, it in turn gives off heat rays, the wavelengths of which are longer than the original radiation.

Absorption, light: The light intensity striking a surface which is neither transmitted or reflected. The exact mechanism of light absorption will depend on the molecules in the substance. Dark substances will obviously have greater absorption than clear ones.

Absorption spectrum: Atoms and molecules in a gas absorb as well as emit light. When the fraction of light intensity absorbed is plotted as a function of wavelength, the result is an absorption spectrum. An absorption spectrum looks like an ordinary spectrum, except there tends to be **dark** lines instead of bright **lines.**

Acceleration: The rate of change of velocity. If an automobile goes from 80 km/hr (about 50 mph) to 100 km/hr (about 62 mph) in 3 seconds, its acceleration is then $(20 \text{ km/hr})/(3 \text{ sec}) = 1.85 \text{ m/sec}^2$.

Acoustic velocity meter: An instrument using the principles of sound to measure the velocity of rivers and sewage pipe fluids.

Alpha particle: This particle, found as a product of radioactivity, contains two protons, each of electrical change 1 unit, and two neutrons, each electrically neutral. Its mass is about 7,000 times that of an electron.

Amplitude: Deflection of a spring.

Angstrom (Å): 10^{-10} meters or one ten-billionth of a meter. A micron is then 10,000 angstroms. If the area of the earth corresponded to one square meter, the area of a fingernail is about the order of a square angstrom. This is an example of "back of the envelope" calculations. The area of a sphere is $4\pi r^2$, and the radius r of the earth is about 6.4×10^6 meters. The area is then 5.1×10^{14} m². If this corresponds to 1 m², then 10^{-20} m² (or 1 square angstrom) corresponds to 5.1×10^{-6} m² or 5.1 mm².

Angular momentum: Momentum is defined as the mass of a body times its velocity or mv. Its units are then g-cm/sec. This velocity implies motion in a straight line. For rotating motion angular momentum can be defined. Angular velocity has units of radians of angle per second or sec^{-1}. The unit of mass times that of angular velocity, in analogy to those of ordinary momentum, is g/sec. A unit of distance is lacking. Take distance from the center of rotation to the body being considered. When we multiply by this distance R, the angular momentum is then $m\omega R$. Consider a clock with a minute hand of a 5-cm radius. This hand has an angular velocity of $0.00174\ sec^{-1}$. An ant with a mass of 1 g at the end of the minute hand has an angular momentum of $1 \times 0.00174 \times 5 = 0.0087$ g-cm/sec.

Angular velocity: The rate of change of angle of a rotating body. Consider the minute hand of a clock with a 5-cm radius. The hand moves 360 degrees in an hour. The angular velocity is usually expressed in radians per second, where the radian is 57.3°. There are 2π radians in 360 degrees. The angular velocity of the minute hand is then 2π radians/3,600 seconds = 0.00174 radians/sec.

Atmosphere: A unit of pressure approximately equal to the average pressure of air on a body at sea level. In more familiar units, it equals 10.1 Newtons (of force) per square centimeter.

Atomic absorption: In a way, atomic absorption is the reverse of emission spectroscopy. White light, containing all wavelengths, is passed through an unknown gas. Depending on the types of atoms present, particular wavelengths of light will be absorbed. When the light through the gas passes through a prism, dark lines appear. The wavelength of each line identifies a particular type of atom.

Background: This is a term used in physical measurements to indicate the effects of normal conditions. For example, in radioactivity measurements the earth and air have small concentrations of natural radioactivity. In outdoor sound measurements, the wind contributes a certain level even if there are no aircraft and cars around.

Bandwidth: The spread in wavelength (or frequency) in a spectral line or other source of light.

Beer-Lambert law: The fraction of light intensity absorbed per unit distance travelled is a constant. The constant depends on what the light is moving through. Suppose 10% of the light intensity is absorbed for every meter travelled. With an initial light unit intensity, after 1 meter the level would be 0.90 units, after 2 meters, 0.81 (= 0.9 x 0.9), and so on. This could be called an exponential decrease, since the change in a distance is proportional to the light level at the beginning of that distance.

Bel: A unit of sound power level or sound intensity level equal to 10 decibels.

Beta particle (or ray): An electron which is emitted from an atomic nucleus. It has properties of both a particle and a ray.

Binding energy: Consider an electron near the surface of a metal. It takes energy to reach in past the other electrons and nuclei in the metal to remove it. This energy is called the binding energy.

Binomial Theorem: An important mathematical concept first used by Newton. A term like $(1 + a)^n$, where a is small compared to 1 and n is an integer, is often encountered. The binomial theorem says the value of this term is $1 + na + n(n-1)^2/2 + \ldots$, where the dots denote higher terms.

Biochemical oxygen demand (BOD): A measure of the oxygen consumed by biological processes that break down organic matter in impure water.

Bioluminescence: The ability of certain forms of life to give off light. The firefly of a summer evening is perhaps the most common example, but certain types of algae can make whole lakes glow.

Biosphere: Roughly defined is that part of the earth and its atmosphere in which living creatures exist. To be more specific, the biosphere extends from a few meters below the surface of the earth to the highest altitude at which birds fly (a few thousand meters).

Black body: A perfect absorber and emitter of radiation. True black bodies do not exist, but a dull black object comes close. A white or shiny body is a poor absorber or emitter of radiation. Perhaps the first test of this statement was when Ben Franklin placed small squares of different colored cloth on snow. The sunlight was absorbed most strongly by black cloth, which melted the snow underneath. The same absorption property also applies to radiation emission.

Boltzmann's constant: Used in many heat and thermodynamic equations. Its value is 1.38×10^{-23} joules per degree Celsius.

Boltzmann's ratio: Relates the number of electrons in a body to their energy levels and temperature.

Brightness temperature: A heated piece of metal changes from red-hot to orange-hot and, finally, to white-hot. Its temperature can be estimated by its color. In turn, its color is a measure of the frequency of electromagnetic radiation being emitted, since visible light is part of this radiation. Radiation is then a measure of temperature.

Buoyancy force: The upward force exerted on a body floating in a liquid. It equals the mass of the displaced liquid times the gravitational constant g. For example, a 70-kg (154-lb) person who just barely floats will displace about 0.07 m^3 of water. The mass of this water is, in turn, about 70 kg. The buoyancy force is then 70 kg x 9.8 m/sec^2 = 686 Newtons.

Calorie (cal): The amount of heat needed to raise the temperature of 1 g of water 1°C. This is a unit of energy like joules. For a long time, heat energy was thought to be different from mechanical or potential energy, so it was given its own units. One calorie equals 4.18 joules. Calories used in nutrition are kilocalories.

Capacitance: The ability to hold electric charge. This definition is in analogy to the ordinary capacity, the ability to hold a fluid.

Capacitor: A device that stores electric charge. In drawing an electrical circuit, a capacitor is represented by two parallel lines perpendicular to the line of current flow.

Centrifuge: A rapidly rotating bowl-shaped machine used for separating substances according to their density.

Centripetal force: The force that produces uniform circular motion. Its direction is along the radius towards the center of the circle.

Characteristic (or natural) frequency: Of an oscillating body, such as a string fixed at both ends; based on its length, the force or tension applied to it, and its mass per unit length. There is generally a basic characteristic frequency as well as higher frequencies of vibration. These are given by twice, three times, four times . . . the basic frequency. The string can then, in principle, have a number of characteristic or resonant frequencies.

Chlorophyll: A green pigment found in most plants.

Coherence: When two or more photons of light with the same wavelength travel in the same direction and have the same phase, they are called coherent. This light has properties different from the incoherent light seen every day.

Colorimetric method: Used to detect pollutants in water by determining the presence of certain ions by color.

Components of vectors: A vector can be defined as a quantity with a direction attached to it. A vector may be heading east, but its effects towards the southeast may be needed. Suppose the vector heading east is $\sqrt{2}$ units long. If a series of vectors are connected, heads to tails, in sequence, the resultant vector is the sum of all the vectors. A vector can then be divided into component vectors if the heads and tails are connected appropriately. In the case under consideration, the eastwards vector can be broken up into one of length 1 in a southeasterly direction and another of the same length in a northeasterly direction.

Conductivity: The ease of passing current through a substance; the inverse of the resistivity. For example, the conductivity of a metal is much higher than that of air. The units of conductivity are 1/ohm-cm.

Convection: Transfer of heat by motion of a fluid, such as air.

Correlation: Exists when there is at least a mathematical relationship between two or more quantities. The simplest way to show this relationship is with a straight line on graph paper. A field of statistical theory is concerned with measuring the degree of correlation between quantities. Of course, some correlations can be false.

Cross section: In collisions between particles and atoms, the cross section is the "target area" that each atom presents to the incoming particle. This cross section is not necessarily related to the physical dimensions of the atom. For example, uranium-235 has a cross section for fission by slow-moving neutrons which is about 100,000 times that of uranium-238, even though the physical dimensions of the two isotopes are the same to within a few percent. The cross section depends on the structure as well as the size of the nucleus.

Curie (Ci): A unit of radioactivity, comprising 3.7×10^{10} disintegrations per second. This number was once thought to be the number of disintegrations from one gram of radium. The unit takes no account of whether the rays given off are alpha, beta or gamma.

Current: The rate at which electrons pass a fixed point in a wire. When one coulomb—a unit of electrical charge—passes in one second, the current is defined as one ampere. *E.g.,* the current passing through an ordinary 100-watt light bulb is just under one ampere.

Decay constant: A measure of the decay rate of a radioactive substance. If r is the decay constant, N_0 is the initial number of atoms present, and N is the number remaining at time t, then $N = N_0 e^{-rt}$.

Decay, radioactive: Certain atoms change their state by emitting particles spontaneously. More precisely, their nuclei give off these particles. The particles can be classified as alpha, beta or gamma. Probably the best-known radioactive elements are uranium and radium. The rate at which atoms decay varies tremendously from one element to another. In addition to spontaneous decay, some elements can be made radioactive using bombardment with subatomic particles.

Decibel (dB): A unit of sound power level or sound intensity level, defined by equations relating power generated or sound intensities to reference levels.

Deuterium: An isotope of hydrogen with a nucleus of one proton and one neutron instead of the ordinary solitary proton. It is written as D instead of H.

Dielectric constant: Related to the concept of capacitance, or the ability of a substance to hold electric charge. The greater this constant, the more charge a body can hold.

Diffraction of sound: Occurs when sound does not travel in a straight line, but "curves" around an opening. The same phenomenon occurs in light, in the well-known pinhole experiment. Light shines on a metal sheet with a tiny hole in it. If the light travelled only in a straight line, it could be seen only directly opposite the pinhole. In fact, it can be seen from many angles, showing that diffraction or bending has occurred. The amount of diffraction depends on the wavelength, and sound is much more easily diffracted than light.

Disintegration, radioactive: See Decay, radioactive.

Dyne: A force that will give a mass of one gram an acceleration of one centimeter per second per second. For example, consider a one-gram mass at rest, with this applied acceleration. After the first second, its velocity is one centimeter per second; after the second, two centimeters per second, and so on.

Electric field: If a small quantity of charge experiences a force, there is an electric field in the vicinity. Fields can be constant or change over time. The most common changing field varies as $\sin \omega t$, where ω is the frequency and t is the time.

Electrochemical potential: A measure of a metal's tendency to form ions in a liquid.

Electromagnetic radiation: Occurs whenever electric charges are accelerated. This radiation is usually defined in terms of its frequency (or wavelength). Electromagnetic radiation is usually classified into (a) infrared radiation, from hot bodies like the sun, with high frequency and short wavelength; (b) visible light; (c) ultraviolet radiation from electric arcs, with wavelengths about one-thousandth of that of infrared; (d) X-rays, from electrons bombarding atoms, with short wavelengths, less than a millionth of a millimeter; and (e) gamma rays, from radioactive atoms, with wavelengths of the order of a billionth of a millimeter. The same physical laws apply to all five types of radiation.

Electron: A subnuclear particle with an electrical charge of -1 unit and a mass of 9.1×10^{-31} kg. The way electrons interact in their orbits around the nucleus forms the basis for all chemistry.

Electron cloud: Electrons are sometimes thought of as circling an atom's nucleus in well-defined orbits. In reality, the electrons "wander" around these orbits, so there is a cloud of electrons, rather than a set of tracks, around the nucleus.

Electron volt: A unit of energy which equals 1.6×10^{-19} joules. This is the energy that an electron receives when it passes through a potential difference of one volt.

Electrostatics: The study of electric charges at rest, in contrast to electrodynamics which studies them in motion.

Emission spectroscopy: If an electric current is passed through a gas, light is emitted. When this light is passed through a prism, it becomes divided into separate wavelengths or lines. Since light has been emitted from the atoms, the process is called emission spectroscopy. Each element has characteristic lines; for example, sodium has bright yellow lines which immediately indicate its presence. The brighter the yellow lines, the more sodium is in the air.

Energy: To use the standard definition, it is the ability to do *work*. In turn, work is defined as the force applied to an object times the distance it is moved. For example, to lift a stone from the ground requires work. The force applied upwards is slightly greater than the force of gravity. If it were equal, the stone would never leave the ground. The distance the stone moves is the height to which it is raised. Since both the force and the distance are known, the work done is also known. When something has done work, it is presumed that it had the ability to do it. Every time work is done, energy was used to perform it. There are many types of energy—chemical, thermal, atomic, potential and kinetic.

Energy level diagram: This shows the energies which electrons in an atom (or molecule) can have as well as the allowed transitions from one level to another.

Equilibrium: All forces in a particular situation are equal and opposite. For example, the force exerted on the bottom of a chair is equal and opposite to that it exerts back.

Excited state: Electrons circling a nucleus have normal orbits. If they have more energy than usual, they travel in orbits which are generally farther away from the nucleus. The electrons are then termed to be in an excited state.

Exponential decay: The inverse of exponential growth. The *decrease* in a short time of a quantity is proportional to the quantity at a particular time. The curve of exponential decay can drop off rapidly or slowly. An example of this phenomenon would be the purchasing power of money under conditions of inflation.

Exponential growth: The increase in a quantity over a period of time is proportional to the amount at the beginning of the period. A bank deposit shows exponential growth if no withdrawals are made. The interest paid is proportional to the amount of deposit, so the money grows exponentially. The growth would be slow at only 1% interest, fast at 10%. Exponential growth is not necessarily fast or slow.

Fallout: When aboveground atomic explosions are set off, the debris of the bomb itself and surrounding earth are sent high into the atmosphere. Part is innocuous, but most is radioactive to a least some degree. The particles gradually drift back to earth through the action of wind, rain and gravity. This fallout can effect humans adversely if food exposed to radioactivity is eaten.

Fission products: Most radioactive decay occurs when an alpha, beta or gamma particle is emitted from the nucleus. These particles are much less sizable than the nucleus. In the case of uranium which has been struck by a neutron, however, the nucleus can break up into approximately equal portions. These elements are called the fission products, and are quite different from those released in normal radioactivity.

Flashtube: In an electrical storm, charge builds up in a cloud (or the earth) until the voltage between the sky and the ground becomes extremely high. A lightening bolt then leaps across the gap. The same thing takes place, on a somewhat smaller scale, in a flashtube. Charge is supplied from a power outlet and is allowed to build up in a capacitor until it flashes through a coil. Electronic flashtubes for cameras work on the same principle.

Flywheels: Any body which rotates and thus can store energy.

Force: The mass times the acceleration. Consider an automobile of 1,500 kg. If it can go from a standing start to 100 km/hr (about 62 mph) in 8 sec, its acceleration is:

$$\frac{100 \text{ km/hr}}{8 \text{ sec}} = 3.47 \text{ m/sec}^2$$

Its force of acceleration is then 5,210 kg-m/sec^2.

Gamma ray (or particle): A photon or "chunk" of light emitted from the nucleus. Its wavelength is generally less than 10 Å and so its frequency is greater than 10^{17} Hz.

Gas chromatography: Polluted water, which in the case under consideration may contain pesticides, is put into a tube with absorbent material. Then a gas is passed through the tube. The gas molecules move particular types of pollutants into different zones along the tube, and based on this motion the pollutants can be identified.

Gravimetry: The process of measuring the density of different substances. In the case of oils in or on water, this method has the advantages of low cost and speed.

Half-life: If radioactive atoms disintegrate, eventually the original atoms will all be transformed. A useful measure of how fast they disintegrate is the half-life, or the time required for half to be transformed. Half-lives very from well under a trillionth of a second (very fast disintegration) to billions of years (very slow).

Heat: A measure of the thermal energy in a body. To determine the quantity of heat involved in a process, the temperature changes are measured and formulas relating the two are used to calculate the heat transferred.

Heat budget: Add all the heat being added to a body. Then do the same for all heat leaving. If the first sum is greater than the second, the body will rise in temperature. If the second sum is greater, the temperature will fall. The process of calculation is called the heat budget.

Hertz (Hz): One cycle per second.

Hologram: A reconstruction of an object in three dimensions using the properties of laser light. In effect, a hologram takes a picture, but displays it so the side and rear can be seen.

Incidence and reflection law of optics: If light strikes a mirror at an angle of 30° from the horizontal line to the right, it bounces off at an angle of 30° from the horizontal line to the left. The same law holds for all possible angles. In other words, the angle of reflection equals the angle of incidence.

Index of refraction: The ratio of the speed of light in a vacuum to its speed in the substance being considered. Air has an index of refraction close to that of a vacuum, so it is sometimes used for comparison. Glass usually has a refraction index of about 1.5.

Inertia: A mathematical expression of the tendency all bodies have to resist motion, or to continue in motion if already moving. For the kinetic energy of bodies moving in a straight line, the inertia corresponds to the body's mass. For rotating bodies, the inertia—technically called the moment of inertia—depends on the shape of the body and the axis around which it is rotating.

Infrared radiation: Refers to electromagnetic radiation longer than the visible red in wavelength. These wavelengths generally run from 8×10^{-4} mm (or 8,000 Å) to about 3 mm. There is no exact dividing line. Infrared radiation is invisible to humans, but may be visible to certain insects and animals.

Integration: In calculus, comprises rules for finding the areas under curves of algebraic functions.

Interference: Occurs when waves of the same wavelength cross each other. Depending on their phase, there can be constructive interference or incoherence. The first-named occurs when the waves are completely in phase, and the second when they are completely out of phase. An analogy to this phenomenon is tugging a rope with one end attached to a wall. The center tends to move comparatively little, because the generated waves and those reflected off the wall are usually out of phase. Interference prevents a strong wave structure in the rope's center.

Ion concentration: If an atom has more electrons than it does normally when it is electrically neutral, it is called an ion. The atom is represented by its regular symbol with one or more minus signs attached. (The atom can also have less electrons than usual, in which case plus signs are attached). A measure of a liquid's acidity is the concentration of ions of a particular type in that liquid.

Isotope: A combination of neutrons and protons is called a nuclide. The number of protons is called the atomic number, but an element can have varying numbers of neutrons. Several nuclides with the same number of protons but different numbers of neutrons are called isotopes of an element. The sum of the protons and neutrons is called the mass number. As an example, consider the isotope U^{235}_{92}. The letter denotes the element uranium, the lower figure is the atomic number, and upper is the mass number. The number of neutrons is then 235 - 92 = 143. Another isotope of uranium is U^{233}_{92}. Isotopes are sometimes written in the form "uranium-235."

Joule (J): A unit of work which equals one newton-meter, or one kg-m^2/sec^2. A joule also equals 10^7 ergs, a much smaller unit. 4.18 joules equals one calorie.

Kilo- : A prefix denoting 1,000, or 10^3.

Kinetic energy: Energy stored in a body due to its motion. The kinetic energy is usually easier to calculate for bodies moving in a straight line or rotating.

Laser: An acronym or shortening of Light Amplification by Stimulated Emission of Radiation. Just as a radio amplifies electromagnetic signals in the air and turns them into sound, lasers amplify or multiply light.

Latent heat: The heat energy required to warm (or cool) a body above (or below) its change of phase temperature. The temperature does not change as latent heat is added or subtracted. For water, there are two latent heats: That of fusion, which occurs when ice melts or freezes, is 80 cal/g; that of vaporization, when water evaporates or condenses, is 540 cal/g. Latent heat exists because it takes extra energy to loosen (or strengthen) the bonds between molecules when a substance changes phase.

Liter: A unit of volume; equals 1,000 cm^3.

Logarithms: A method of calculation which allows addition and subtraction to be substituted for multiplication and division. Consider multiplying 3.007 by 2.973. The logarithms of these numbers are 0.47813 and 0.47319. Added together, this is 0.95132. The antilogarithm of this, or the number to which the logarithm corresponds, is 8.9395. The logarithms are added instead of multiplied, because of the rule log (ab) = log a + log b, where a and b are any two numbers. Similarly, log (a/b) = log a - log b.
The antilogarithm of a number raises it to the power 10. For example, the antilogarithm of 0.5 is $10^{0.5}$ = 0.3162. One useful quantity to remember is that the logarithm of 2 is very close to 0.3.

Log-log graph paper: This considers the logarithms of experimental values, rather than the numbers themselves. For example, an original set of values along the y-axis (or vertical axis) might have been 10, 20, 30, 40 . . . . 100. The logarithms of these numbers are 1, 1.30, 1.48, 1.60 . . . . 2. Using ordinary graph paper, 30 would be the same distance (10) from 20 as 20 is from 10. In logarithmic paper, the number corresponding to 30 would be 1.48, only 0.18 units from the number corresponding to 20, which is 1.30. The latter number is 0.30 units from the number corresponding to 10, which is 1. The distance is now different. The upper end of the scale is compressed. Suppose a law is of the form y = kx^a, where k and a are constants. Experimental measurements which obey this law will be a straight line on log-log graph paper.

Loudness: Not a physical measure like decibels or watts, loudness is considered in psychological and physiological terms.

Mass spectroscopy: Pollutant molecules have different masses. If their masses are known the pollutants can often be identified. The procedure is based on the fact that a charged particle moving in a straight line will curve when it enters a magnetic field. The amount of curvature depends on the ratio of its electric charge to its mass. If the particle is highly charged and light, it will curve sharply, with a curvature of small radius. If the particle is heavy with little charge, it will hardly curve at all. By measuring the radius of curvature, the ratio of charge to mass of the pollutant molecules can be found. Since the value of the charge is often known, the mass can be found and the particular pollutant identified.

Mega-: A prefix denoting million, or 10^6.

Mho: A unit of electrical conductance, the opposite of resistance. It is defined to equal 1/ohm.

Micro- : Denotes 10^{-6}, or one millionth.

Microdensitomer: A device that converts light into electrical signals. The instrument moves over a photograph or a piece of paper, and produces a high voltage when a dark spot is crossed and a low voltage when it passes over whiteness. The direction of the voltages can be reversed if necessary.

Micron (μ): One millionth of a meter. A sheet of paper is usually around 100 microns thick.

Milli- : A prefix denoting one thousandth, or 10^{-3}.

Monochromaticity: In Latin, "one color." The purest colors are those emitted in spectra, but even these are not perfectly pure. If they were, they would have exactly one wavelength (or frequency) without any spread. Laser light comes the closest to monochromaticity, with very narrow spreads.

Natural frequency: See Characteristic frequency.

Neutron: A subatomic particle with no electric charge and a mass about 1,800 times that of an electron. When it moves slowly enough, it can make uranium-235 fission or break apart.

Neutron activation analysis: A method of identifying substances by bombardment with neutrons. The neutrons disintegrate the nuclei, causing them to produce particles with energy characteristic of particular elements.

Newton: A unit of force, it equals 1 kg-m/sec^2.

Noisiness: See Loudness.

Nucleus: The core of an atom; occupies about one ten-thousandth of its diameter. Composed of protons and neutrons, it has a far greater mass than the rest of the atom.

N-wave: The change in air pressure due to a sonic boom. The pressure rises sharply, declines slowly to less than normal, and then rises rapidly to normal. The entire process is over in a few thousandths of a second. When the pressure is graphed, it looks like the letter N.

Ohm: A unit of electrical resistance. As an example, the resistance of an ordinary 100-W light bulb is about 120 ohms.

Optics: The study of light visible to both the eyes and scientific instruments.

Order of magnitude: Generally taken, in physics, to mean approximately ten. For example, the number of weeks in a year is an order of magnitude less than the number of days. The idea can be used to separate out quantities by their approximate sizes in preliminary calculations, and so help to solve complicated physical problems.

Organic compound: One which contains carbon atoms. The term was originally applied to compounds in living organisms.

pH: A measure of the acidity of a substance, primarily a chemical concept.

Phase, photon: Light can be thought of as vibrating. An analogy on a larger scale is the so-called simple harmonic motion of a skipping rope, one end of which is attached to a wall. The phase of a point in the rope would be measured by its vertical position compared to its position in a taut rope. The part of the rope that is farthest from the ground is completely out of phase with the part that is lowest. Photons of light also have a measurable phase.

Photoelectric effect: Occurs when light of certain wavelengths (usually ultraviolet) strikes a metal. Electrons, or negative electric charges, are ejected from its surface.

Photoelectron spectroscopy: Every material will have a characteristic binding energy if it has a photoelectric effect. Photoelectron spectroscopy finds these energies and uses them to identify substances.

Photomultiplier tube: Uses the photoelectric effect (see above) to transform light into electrical signals. The signal produced by the light is usually weak and is amplified or multiplied electronically in the tube.

Photon: A unit, or quantum, of electromagnetic radiation or light, which acts both like a wave and a particle.

Photosensitivity: Any material which exhibits an effect due to light shining on it can be called photosensitive. Selenium, the metal which is the basis for xerocopies, is photosensitive, as are most plants.

Pico-: A prefix for 10^{-12}, or a millionth of a millionth.

Piezoelectricity: When a slight pressure is applied to certain crystals, tiny voltages are produced. This effect is known as piezoelectricity. It is used in such everyday devices as microphones and phonograph pickups. Piezoelectricity is a simple way to translate pressure into an electrical signal.

Planck's constant: Relates the energy of a photon to its frequency of vibration. Its value is 6.63×10^{-34} J-sec.

Planck's law: Discovered around the turn of the century, Planck's law relates the energy given off by a black body (or perfect radiator) to its wavelength λ and temperature. The equation in the text should have a factor $(1/\lambda)^5$ multiplying the right-hand side to be completely factual, but since the discussion does not concern wavelength, the omission is permissible. The extra factor is in the constant K_1. The exact shape of the curve depends on wavelength and temperature, but is approximately bell-shaped at high temperatures when energy is plotted as a function of wavelength.

Polarography: A method of measuring the relationship between voltage and current in an electrical circuit containing liquid solutions. The voltage can be defined as the differences in energy between different groups of electrons in the circuit. The current is a measure of how many electrons per second pass a given point in the wire. The importance of polarography lies in the fact that pollutants have characteristic graphs of voltage vs current, and these graphs can be used to identify them.

Population inversion: When there are more electrons in matter at a particular energy level than expected on the basis of Boltzmann's ratio, and other energy levels have the expected number, there is an inverted population.

Porosity: Here means a measure of how agitated a suspension of solids in water has become. In the usual sense, it is indicative of how many pores there are in a solid. For example, porous rock can absorb a lot of water or oil.

Potential difference: See voltage.

Potential energy: Due to the relative positions of the parts of a system of bodies. A stone raised in the air has potential energy with respect to the earth below.

Power: The ability to do work, or the rate of doing work.

Radioactive tagging: For many physical and biological experiments, small quantities of radioactive materials can be passed through the system. Since radioactivity is simple to detect, a better understanding of how the system works can be gained. For example, chemically measuring how potassium moves through plants is time-consuming. Doing the same measurements using a radioactive isotope of potassium is much simpler.

Radiometer: In principle, any device which can detect radiation. As a result, an ordinary radio or TV is a radiometer as well as the more sophisticated apparatus in a physicist's lab. One of the first radiometers was invented by Crookes, and consisted of small pieces of mica mounted on two arms at right angles to each other. One side of each mica piece was blackened. The radiation pressure of a light bulb on the dark mica is different from that on the unpainted mica, and the pressure is great enough to turn the arms. The device is still seen at science fairs.

Reference level: Used in computations of sound power or intensity levels, as a denominator in the fraction containing the sound intensity or power as numerator. For power levels, the reference level is 10^{-12} W; for intensity levels, it is 10^{-12} W/m^2.

Reflectance (or reflectivity): The fraction of light intensity which is reflected at a surface. The reflectance of a mirror will generally be close to 1.

Reflection: When a light ray strikes a surface, there can be reflection or light leaving the surface in a definite direction.

Rem: A unit of radiological dose to biological tissue. It is calculated by multiplying the absorbed dose in rads by the relative biological effectiveness (RBE) of the particle doing the damage. For example, X- and gamma rays have an RBE of 1; fast-moving neutrons 10, and alpha particles 10-20. The RBE takes into account the energy and mass of the particle being absorbed.

Remote sensing: Includes all methods whereby physical properties are measured from a distance. In common practice, special film is exposed, either from aircraft or earth satellites, and then interpreted on the basis of the colors or dark sections which are seen. The film gives an indication of the wavelengths being emitted from the ground or water below.

Resistivity of a substance: Can be roughly defined as how difficult it is to have electrical current pass through it taking into account its physical dimensions. Its units are ohm-centimeters.

Resonant frequency: All vibrating systems have one or more resonant frequencies, which depend on what they are made of, their dimensions and other factors. If the system has a force applied to it at that frequency, the vibrations will be much greater than at slightly different frequencies. For example, a car may be driven at 100 km/hr and feel as if the wheels were about to fall off, yet there may be little or no effect at 98 or 102 km/hr. The frequency of vibration at 100 is clearly resonant.

Scattering, heat radiation: Occurs when radiation waves encounter dust particles in the air. Since these waves also can be thought of as particles, they can be visualized as "bouncing" off the dust in all directions. The radiation is then prevented from travelling in a straight line.

Scattering, light: A type of diffuse reflection. If light strikes a rough surface and there are no flat spots much bigger than a wavelength, the light will scatter.

Scientific notation: Useful for very large or very small numbers. The number in question is divided into two parts. The digits at the beginning are written with a decimal place. This is followed by 10 raised to the appropriate power. The same principle applies to small quantities, where the power of ten is one more than the number of zeros before the first nonzero digit. Thus, 0.00027 is 2.7×10^{-4}.

Semilogarithmic graph paper: Similar to log-log graph paper, except that only the y-axis has a logarithmic scale. Equations of the form $y = k\, e^{ax}$, where k and a are constants, form straight lines on this type of paper. On ordinary paper they would be curves.

Sigmoid curve: Many natural phenomena can be described by an s-shaped or sigmoid curve. In this curve, the quantity being considered rises slowly at first, rapidly in the middle portion and then slowly until it reaches a levelling-off point. The x-variable is usually time. For example, consider the earth's population. For thousands of years, it grew slowly as man coped none too successfully with nature. In the past few centuries, it has grown rapidly due to advances in science and medicine. It seems clear that this growth will level off as resources become depleted.

Simple harmonic motion: Occurs when the restoring force on a displaced body is proportional to the distance it has moved. For example, consider a spring which has been stretched 0.1 mm and then 0.2 mm. The force trying to bring it back to its normal position for the second case will be twice that of the first.

Sonic boom: Occurs when the source of sound waves moves faster than the waves themselves can travel. Because the velocity of sound in air varies with the density of air, which in turn varies with the distance from the ground, the occurrence of a sonic boom depends on the height of the aircraft above sea level. The sonic boom travels in a cone, the angle of which depends on the velocity of the aircraft. If the velocity is just above the velocity of sound, the cone will be wide; if the velocity is much above that of sound, the cone will be narrow.

Specific conductance: Indicates how easily electricity can pass through water containing different concentrations of other materials, such as pollutants.

Specific gravity: Proportional to the density of a substance. Since it is defined as a ratio with respect to water, it is a pure number without any units.

Specific heat: Suppose the same amount of heat energy is absorbed by the same mass of aluminum and steel. Although the heat energies are equal, the resulting change in temperature is not. The temperature of the iron rises twice as fast as that of the aluminum. The property of matter which quantitatively describes these differences is the specific heat, defined as the number of calories needed to raise one gram one degree C. Water has a higher specific heat than any common metal, with a value of 1.00 cal/g-°C.

Spectrum: A hot solid, liquid or gas emits a spectrum that depends only on the material and its temperature. It thus is a reliable fingerprint. The spectrum is a plot of light intensity vs wavelength (or frequency). Spectra for gases are a series of bright lines, and those for solids and liquids are usually continuous, or lines overlapping each other.

Spring constant (or stiffness): A measure of how resistant a spring is to pulling. Car springs, although built somewhat differently, will have a higher value of this constant than those of a set of household scales.

Standard deviation: Suppose there are many measurements of a quantity presumed to be similar, like the diameter of peas in a pod. If the number of readings for each diameter were plotted, a bell-shaped curve would probably result, with a few small and large peas and most clustered around the average size. About two-thirds of all measurements fall in the range spanned by the standard deviation, a measure of the spread. If the spread is small, the standard deviation is low; if the spread is large, it is high. The quantity is used in many statistical calculations.

Standing waves: Occur when the ordinary travelling or moving waves "reflect" from the ends of a string and produce other waves moving in the opposite direction. They exist when both ends of the string are clamped or fixed. The waves are called standing because the energy in the string cannot move past the fixed ends. In other words, it "stands" in the string. In real strings, the standing waves soon die down due to friction in the string and with the air.

Stiffness: See Spring constant.

Stimulated emission of light: Occurs when a substance has an "inverted" population structure of electrons with respect to their energy levels, and photons of the correct wavelength are injected into it. Its properties are different from ordinary light.

Strength: The pressure per unit cross section of a filament of the material.

Stress: See Strength.

Surface tension: Keeps a razor blade floating in liquid rather than allowing it to sink, in the well-known experiment. Its units are force/length.

Thermal conductivity: A measure of how fast heat passes through a substance. For example, suppose there were two houses with walls of equal thickness; one made of glass and the other of asbestos. On a cold day, heat would pass through the glass house much faster. The thermal conductivity of glass is then higher than that of asbestos.

Trace element: Any element which has a very small concentration in a solid, liquid or living organism. Mercury is a trace element which has received considerable publicity in the last decade. While there is no definite cut-off point, a trace element is generally taken to be in the parts per million or billion range.

Transmittance: The fraction of light intensity which is transmitted through a body, such as a pane of glass or a particle of asbestos. For glass, the transmittance will be close to the maximum value of 1 if a light is directly overhead. If the light is to the side or making a small angle with the glass, the transmittance can be close to its minimum, or zero.

Tritium: An isotope of the element hydrogen, one of only two isotopes with its own name (deuterium is the other). Instead of one proton in its nucleus like hydrogen, it has one proton and two neutrons. It is produced in certain types of nuclear reactors. Tritium occurs in negligible amounts in nature.

Turbidity: A measure of how cloudy a liquid is. It can be measured by shining light through the water, and finding how much has been absorbed.

Ultraviolet radiation: Usually has a wavelength between 10 and 4,000 Å. This is shorter than visible light at the blue or violet end, and its name means "beyond violet." Its usual source is electric arcs.

Vapor pressure: In a mixture of gases, the pressure exerted by each gas. In common usage, the term is often applied to water vapor pressure.

Vibration: Small deflections of a body around an equilibrium point. It usually can be described by simple harmonic motion.

Virtual image: This is what is seen in an ordinary mirror. There appears to be someone a meter or so behind the mirror. A discussion of real and virtual images is beyond the scope of this book, but a rule of thumb is that real images can be shown on a screen; virtual ones cannot.

Viscosity: Resistance to fluid flow; lack of slipperiness. The viscosity of molasses in January is high.

Voltage: A measure of electrical potential difference. Just as there is a gravitational potential difference between a book on a table and the same book on the floor, electrical potential differences also exist. If charge is moved from one potential to another, work is done. The voltage is a measure of how much work is done per unit charge. The most common voltage is that of household electricity, 110 V.

Waste heat: Occurs in any engine using heat. The efficiency of a heat engine, which can be defined as any device getting mechanical work from heat, can be defined as $(T_1 - T_2)/T_1$, where T_1 is the absolute temperature of the heat input and T_2 is the absolute temperature of the heat output. If there was a steam engine being heated at 300°C with an exhaust temperature of 100°C (the temperature of steam), the efficiency would be $(573 - 373)/573 = 35\%$. If 1 joule of energy were expended, at most 0.35 joules of mechanical work could be used. Most of the rest of the energy would be waste heat.

Watt (W): A unit of power, it equals one joule/second, or one $(kg \cdot m^2)/sec^3$.

Wavelength: Electromagnetic radiation can be thought of as varying in intensity over time, just like the heights of ocean waves or skipping ropes. The wavelength is then defined as the distance between the crests.

Weighting: A mathematical or physical method of giving more prominence to one range of a variable than another. For example, economic data may be weighted by the amount each person earns, with those having the largest salaries being assigned the largest weights. In sound, measuring instruments are weighted so their response is comparable to that of the ear.

Wilson cloud chamber: Consider a box with a piston that can push in or out, just like the piston in a gasoline engine. Suppose there is enough water vapor in the enclosed air to make it saturated. Any more water vapor and drops would run down the walls. If the piston is pushed inwards, the air becomes supersaturated, and any slight disturbance, such as an atomic particle, can produce "condensation centers," or droplets of water. These droplets can be photographed to measure the number of particles or their direction of travel.

Work: See Energy.

INDEX

Absolute
  temperature 161
  zero 74,75
Absorption 5,68-71,127,129-133,
    136-139,144,164-166,170
  atomic 7-9
Absorption, sound
  See Insulation, sound
Absorptivity 69,70
Acceleration 23,28,35,48
Acid, mine 105-107
Acid rain 98-100
Acoustic velocity meter 49-53
Air pollution 3-7,22,24,53,91-102,
    115-117,127,128,142-148,
    165,168-170,184,185
Algae 126,131
Angular velocity 22,34
Approximation
  See Order of magnitude
Asbestos 127,128
Atmosphere 2,10-13,91-93,144,
    176-178
Atom 173
Automation 6-8,10
Automobile 10,46-48,169,170

Bacteria 54,55,149,185
Bandwidth
  See Monochromaticity
Beer-Lambert Law 129,130,144,
    164,165

Bel 42
Beta particles
  See Electrons
Binding energy 146,147
Binomial theorem 96
Biochemical oxygen demand 185-
    187
Bioluminescence 126
Biosphere 11-13
Black body 72
Boltzmann's ratio 160,161
Brightness temperature 108,109
Buoyancy 21,26,28

Calorie 80
Capacitance
  See Capacitor
Capacitor 99,102-105
Carbon dioxide 10-13,16
Centrifuge 34,35
Chlorophyll 126,131-136
Chromatography, gas 8
Climate 10,80-82,133
Colorimetry 8
Conductance 2,4,10,87-93
  Also see Resistance; Resistivity
Conduction, heat 77,81
Conductivity
  See Conductance
Convection 75
Cosmic rays 174
Costs 7,10

Cross section, nuclear   179
Currents, ocean   109,110,126

Decay constant   179
Decibel   41-43,46
Density
   See Gravity, specific
Detectibility, limits of   7,9
Deuterium   181-183
Dielectric constant   4,102,103,105
Differential (calculus)   130,176
Diffraction   59
Digitizing   140
Dust
   See Particulate matter

Electric field   94,116
Electric potential
   See Voltage
Electricity, static   53
Electrochemical potential   111
Electromagnetic radiation
   See particular types, *e.g.,* Gamma rays, Microwaves
Electron   146,147,159,161,179
   volt   160,161,180,184
Electrophoresis   114
Electrostatics   115-117
Energy
   balance   101,102
   binding   146,147
   kinetics   22,146,147
Epilimnion   81
Errors   13-16,52,133,169,170
Eutrophication   135
Evaporation   75-77,81,82
Exponential growth   10,18,102

Fallout   176-178
Fechner's Law   42
Fertilizer   89,90,131
Field, electric   94,116
Filter   94,115-117

Fish   87,88
Fission   176
Flashtube   158,161
Fluorescence   3-6
Flywheel   22-25,38
Food chain   87
Force, centripetal   22,25,35
Frequency, characteristic   53,64,96

Gamma rays   179-190
Graph paper
   log-log   38,164
   semi-log   77,100,134,178,180
Gravimetry   8
Gravity
   force and acceleration of   26-29, 33,35
   specific   34
Gulf Stream   109

Half-life   184
Hearing   43-49
   Also see Noise pollution; Sound
Heat
   budget   74
   latent   81
   specific   75,80
Highways   46,47
Holography   166-168
Hydrogen   160

Inertia   27,28
   moments of   22
Infrared   8,126-128,138-139
Instrumentation   1,3,5-10
Insulation, sound   56-61
Intensity level (sound)   42-49,57,60
Ions   107,111,116,145-148
   specific   7-9
Isotope   14,177,184

Joule   80

Joule's Law   109

Lake Erie   131
Lake Tahoe   133
Lasers
   coherence   157-161,167
   directionality   157-159,164
   frequency   165
   intensity   157,159,164-165
   principles of   157-161
Latent heat   81
Linearity (mathematical)   96,97
Luminescence
   See Fluorescence

Man-rem
   See Rem
Mercury   87,178
Methylmercaptan   141,143
Mho (electrical unit)   88
Microdensitometer   144
Microwaves   107-109
Mining   2,105-107
Momentum   34
Monochromaticity   165
Motion, simple harmonic   31

Neutron activation analysis   7,9,
   178-182,184,189
Newton   21,125,173
Nitrous oxide   168-170
Noise pollution   41,43,61,141
Nuclear cross section   179
Nuclear reactors   13,173-176,190
Nuclear weapons   173-178,190

Ocean   131-137,186
   currents   109,110,126
Odor   140-143
Ohm's Law   101
Oil   2-4,8,25-31,38,87,102-105,
   107-109,126,127,136-141,
   150,163-165,188-190

Optics, geometrical   125
Order of magnitude   25,28,31,96,174
Outfalls   2
Ozone   166

Particulate matter   53,54,74-78,92-
   98,100-102,115-117,127,144
   size   166-168
Pesticides   8,10,150
Petrochemicals   2
pH   2
Phosphoric acid   89,90
Photoelectric effect   146-148
Photoelectron spectroscopy   147
Photons   146,157-159,161
Photosensitivity   150
Physicists   1,11,13,25,31,32,56,63,
   69,89,96
Phytoplankton   132,133
Piezoelectricity   94-98
Planck's Constant   146,160
Planck's Law   72
Polarography   8,111-113
Pollution
   See Air; Noise; Thermal; Water
Population inversion   161
Porosity   36,37
Power plant efficiency   80
Power, sound   41-43,46,47
Precipitator efficiency   116,117
Precipitation electrostatic   115-117
Pressure   24
Protons   179,189

Rad (unit)   186,187
Radiance   136,137
Radiation
   background   174
   thermal   74-77
Radioactivity   1,2,10,13,14,173-190
Radiometer, differential   132,133,
   136
Reactors, nuclear   13,173-176,190
Reflection   127,129,131,162,163,165

Reflectivity 2,108
Rem (unit) 174
Remote sensing 70-73,132-134, 144,145
Reproducibility 6,8
Resistance, electrical 101,103,110
Resistivity 97,106,107
Rivers 2,49-53,67
Runoff, fertilizer 2

San Francisco Bay 133,135
Santa Barbara (oil spill) 138
Scattering 68
Seas
 See Ocean
Seaweed 129-133
Sensitivity 6,8,63
 human 31-33
Sewage 2,34,52,54,55,126,131, 185-188
Shellfish 149
Ships 2,188,190
Simple harmonic motion 31
Sludge 34,35,38,54,55
Smell
 See Odor
Smog 145,148
Smoke
 See Particulate matter
Sonic boom 56
Sonication 54,55
Sound
 insulation 56-61
 intensity level 42-49,57,60
Specific electrodes 8,9
Specific heat 75,80
Specific ions 7-9
Spectrography
 See Spectroscopy
Spectroscopy 7-9
 mass 8
 photoelectron 147
Spectrum 4,48,126,132,141,156
Standard deviation 133,140,189, 190

Strength (of materials) 24,25
Stress
 See Strength
Strontium-90 177,178
Sulfur in air 145-148,166
Sulfuric acid 98-100,102

Temperature
 absolute 161
 brightness 108,109
 water 70,72-82,136
Tension, surface 4
Thermal map 70-73
Thermal pollution 2,67,68,70-73, 76,127,136
Thermal radiation 74-77
Thermometer 67
Trace elements 7,178,180,184
Tracer, radioactive 185-187
Transmittance 127-129,144,145
Tritium 183,184
Turbidity 2,126,162

Ultraviolet 149,150
Upflow clarification 36,37

Vapor pressure 182
Vibration 31-33,53-55,57,94,95,97
Viruses 149
Viscosity 3,4,30
Vision 125
Voltage 7,71,87,88,115-117

Wastes
 inorganic 2,6
 organic 2,6,8,185
 Also see Pollution
Water
 drinking 36,181-183
 pollution 1-3,6,87-91,105-107, 110-113,126,133,149,161-163, 187,188

Water
   sea   3,4,131-137,149,185
   temperature   70,72-82,136
Waves   3,56,57,126
   standing   54
Weather
   See Climate

Weighting (sound   48,49
Wilson cloud chamber   100

X-rays   174